AF251292

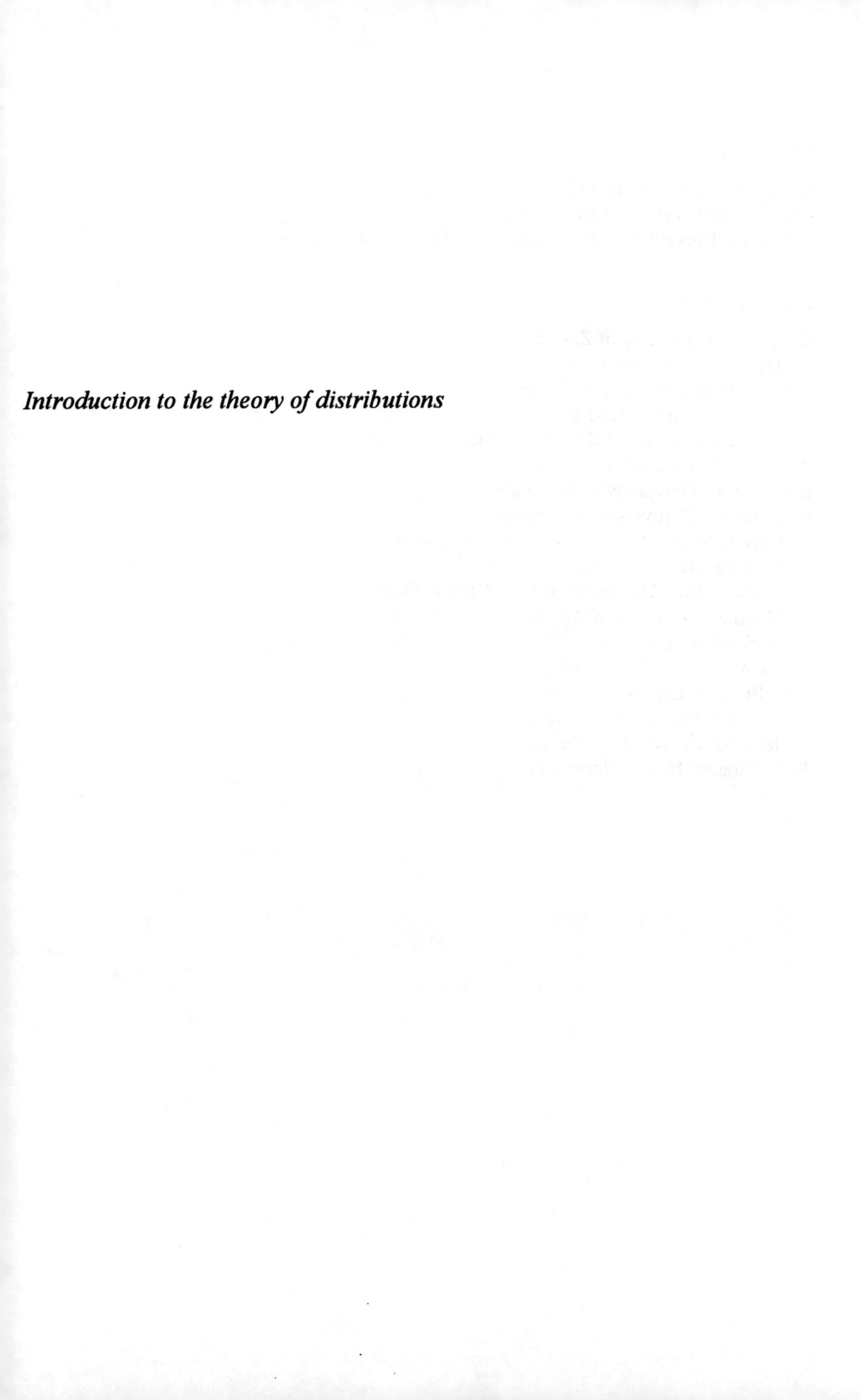

Introduction to the theory of distributions

Introduction to the theory of distributions

J Campos Ferreira

Instituto Superior Técnico, Lisbon

Translated by

J Sousa Pinto

University of Aveiro

and

R F Hoskins

De Montfort University

LONGMAN

Addison Wesley Longman Limited
Edinburgh Gate, Harlow
Essex CM20 2JE, England
and Associated companies throughout the world.

*Published in the United States of America
by Addison Wesley Longman Inc.*

© Addison Wesley Longman Limited 1997

This text is the translation of "Introdução à Teoria das distribuições",
1st edition, published by Calouste Gulbenkian Foundation, Lisboa 1993.

The right of J Campos Ferreira to be identified as author of this
work has been asserted by him in accordance with the Copyright,
Designs and Patents Act 1988.

First published 1997

AMS Subject Classifications: (Main) 46-02, 46-01, 46F99
                            (Subsidiary) 26A24

ISSN 0269-3666

ISBN 0 582 31144 6

British Library Cataloguing in Publication Data

A catalogue record for this book is
available from the British Library

Printed and bound by Bookcraft (Bath) Ltd

M.N.
3-3-98

Contents

Preface to the English Edition

This book presents an introduction to the theory of distributions based on the original and very simple point of view developed by J. Sebastião e Silva (1914-1972). Internationally recognised by many of his contemporaries as the most distinguished Portuguese mathematician of this century, Sebastião e Silva made significant contributions to several widely varied domains of mathematics: algebraic equations, mathematical logic, general topology, etc. However, his most relevant and important work concerns Functional Analysis: the theory of analytic functionals, the theory of the locally connected spaces, differential calculus on locally convex spaces, symbolic calculus and generalised function theory. Concepts which he created and the results which they generated have been the origin of later work by G. Köthe, A. Grothendieck, J. Colombeau and many others.

With regard to generalised functions and more particularly to the theory of distributions of L. Schwartz, the axiomatic approach of Sebastião e Silva surely constitutes one of the simplest and most natural ways in which to introduce the fundamental ideas of the theory. Within this approach the essential concepts and results of the theory become easily accessible, not only to physicists and engineers with no particular specialised mathematical background but also to begining years university students who have already attended a course of elementary differential and integral calculus.

Sebastião e Silva expounded his ideas on the subject in several research papers and in courses delivered in a number of universities but, although he had planned to write a book on distributions, he was never able to do so. Hence, the Portuguese version of the present book, published by the **Fundação Calouste Gulbenkian** in 1993, constitutes the first expository text in which this theory of distributions is treated from the Sebastião e Silva point of view. However the fact that the book was written in Portuguese inevitably made it difficult of access for many potential readers, and the desirability of an English translation became clear.

This present edition is designed to overcome this difficulty and has been produced with the help of two colleagues, specialists in the area of generalised functions, Professor J. Sousa Pinto from the University of Aveiro, Portugal, and Professor R. F. Hoskins now with De Montfort University, England. Professor Sousa Pinto has generously devoted much time and trouble to the preparation of a translation of the complete text from the Portuguese while Professor Hoskins has kindly undertaken a revision of the resulting English version. It is with great pleasure that I take this opportunity to register my deep gratitude to each of these colleagues, without whose stimulus and collaboration the work would never have been completed.

Lisboa, April 1996

(J. Campos Ferreira)

Preface to the Portuguese Edition

This text constitutes the first volume of a larger work devoted to the exposition of the main features of the Laurent Schwartz theory of distributions from the point of view developed by José Sebastião e Silva. In this volume we consider only the case of distributions defined on open subsets of the real line; the study of distributions (both scalar and vectorial) of several variables will be left to a later volume as also will be some applications and complementary aspects of the theory. This allows the basic ideas of Sebastião e Silva to be introduced with the maximum degree of simplicity, and at the same time to outline clearly the path to be followed in the subsequent generalisations.

The exposition is fundamentally inspired by courses given by Sebastião e Silva at the Universities of Lisbon, Oporto and Maryland, together with some of his research papers. Some contributions of the present author to the treatment of values and limits of distributions have been inserted in chapter 2 and appendix A, with some applications in that chapter and in the following. The writing has also been influenced, with regard to certain specific points, by several papers produced by other research workers of the Portuguese school created by Sebastião e Silva.

Prerequisites for the theory developed in this book are quite modest: knowledge of the main subjects usually taught in courses on Mathematical Analysis and Linear Algebra during the first two years of Physics or Engineering Studies in the Portuguese Universities. To ensure such a level of accessibility however, it has been necessary on the one hand to include some items of Analysis not usually dealt with in such courses and, on the other, to omit certain more delicate aspects of the theory of distributions (these being generally less important from the point of view of applications). For example, instead of the proper topological study of spaces of distributions we consider only those aspects which may be treated in terms of specifically defined modes of convergence of sequences. Particular emphasis is also given to the concept of distributions of finite order, this being in fact the main object of study throughout the whole of this volume. Following the usual terminology of Sebastião e Silva we use the term *distribution* to refer to just those objects which Schwartz calls *finite order distributions*, reserving for the general concept of a Schwartz distribution the term *global distribution* (this term actually appearing only in the very last section of the book).

In particular it should be stressed that with respect to integration a knowledge of the Riemann integral is sufficient. The fact that we have avoided, for the moment at least, any acquaintance with more complete theories of integration does involve us in certain limitations. In particular this is the case in connection with the possibility of identifying with distributions certain functions or classes of functions.

The writing of this volume has been achieved during the author's period of sabbatical leave in 1986/87. Since then it has not been possible to pursue further the main project it seemed preferable to publish this first part in a definitive form while waiting for a better opportunity to conclude the full work. For this I have had the support of Fundação Calouste Gulbenkian, for which I wish to express my sincere gratitude.

Lisboa, March 1990

(J. Campos Ferreira)

Introduction

Although there have been many forerunners - Heaviside, Dirac, Leray, Sobolev, etc. - who have made important contributions to the creation of distributions, the first systematic and fully developed study of the theory, published in 1950-51 in the Actualités Scientifiques et Industrielles (n°s 1091 and 1122), is due to the French mathematician Laurent Schwartz.

In the Schwartz approach much of the theory is based on deep concepts and results from Functional Analysis, so that its study requires quite a lot of knowledge of this discipline, and therefore presents difficulties to many potential users. As a result there have since appeared several alternative approaches to distributions, designed to develop the most important aspects of the theory with fewer prerequisites (see [1], [20], [25], etc.). Among such approaches there is the work developed by Sebastião e Silva.

As we will see in the sequel the process by which Sebastião e Silva introduces distributions is both elegant and extremely simple. The whole theory is based on a small number of simple axioms suggested naturally by an intuitive concept of distributions which itself provides a privileged model for them. Within that model distributions are regarded simply as derivatives of ordinary continuous functions, the concept of differentiation being suitably reinterpreted in a generalised sense.

In a non-rigorous sense we may say that the concept of distribution generalises that of function (or, at least, of continuous function) very much in the same way as, for example, the concept of complex number generalises that of real number. One of the essential objects of such generalization is to allow differentiation to be possible in much more general circumstances than those obtained in Classical Analysis. In consequence it becomes possible completely to justify several methods of empirical reasoning and calculation which have been used by physicists and engineers. In particular this is the case with the symbolic or operational calculus of the electrical engineers and with methods used in Quantum Mechanics.

Among the distributions which first appeared within the classical setting we may mention in particular the so-called *"Dirac δ function"*, or as the electrical engineer calls it, the *"unit impulse function"*. Here is one of the most natural ways in which it may be introduced:

Suppose we are given a mass distribution on the real axis $\mathbb{R}$. Then to each bounded interval $\mathbf{I} \subset \mathbb{R}$ there will correspond a nonnegative number $m(\mathbf{I})$ representing the quantity of mass contained in $\mathbf{I}$. In many cases of practical importance there will exist a function ρ, of a single real variable x, which satisfies the equation

$$m(\mathbf{I}) \;=\; \int_{\mathbf{I}} \rho(x)\,dx, \tag{0.1}$$

for each bounded interval $\mathbf{I}$. Then it is easy to see that at each point x at which the function ρ is continuous we have

$$\rho(x) \;=\; \lim_{h \to 0} \frac{m(\mathbf{I}_{x,h})}{h}\,, \tag{0.2}$$

where $\mathbf{I}_{x,h} = [x - h/2, x + h/2]$. As is well known the quantity $\rho(x)$ is called the *density of the mass distribution at the point x*.

Now consider a particularly simple distribution of matter on $\mathbb{R}$ obtained by placing a material point of unit mass at the origin, but with the rest of the real axis left free of matter. In this case we would have $m(\mathbf{I})$ equal to 1 or to 0 according as $0 \in \mathbf{I}$ or as $0 \notin \mathbf{I}$. If such a distribution of mass were to have a density, denoted by $\delta(x)$, then we would have

$$\int_a^b \delta(x)\,dx \;=\; m([a,b]) \;=\; 1,$$

for every a, b such that $a < 0 < b$. Further, if this density function were to satisfy condition (0.2) we would also have

$$\delta(x) \;=\; \lim_{h \to 0} \frac{m([x - h/2, x + h/2])}{h} = \left\{ \begin{array}{ll} 0 & \text{if} \quad x \neq 0 \\ +\infty & \text{if} \quad x = 0. \end{array} \right.$$

Given this situation, and other similar ones, some physicists and engineers have been led to postulate the existence of a function δ, defined on the real axis, which satisfies the conditions

$$\delta(x) = \left\{ \begin{array}{ll} 0 & \text{if} \quad x \neq 0 \\ +\infty & \text{if} \quad x = 0. \end{array} \right. \tag{0.3}$$

together with

$$\int_a^b \delta(x)\,dx \;=\; 1 \tag{0.4}$$

for any real numbers a and b such that $a < 0 < b$.

It is clear that condition (0.3) defines a function on $\mathbb{R}$ with the peculiar property of assuming an infinite value at the origin. This fact is not in itself a novelty; the use of functions assuming values in the extended real number system, $\bar{\mathbb{R}} = \mathbb{R} \cup \{-\infty, +\infty\}$ was already current at the time in certain areas of mathematics, such as measure theory. But condition (0.4) is in direct contradiction with (0.3) under any theory of integration, classical or modern. The function δ defined by (0.3) is certainly Lebesgue integrable, for example, but its integral over any interval of $\mathbb{R}$ is equal to zero since this function differs from the null function at only one point. While we might agree to introduce a new concept of integral which would make conditions (0.3) and (0.4) compatible, this would require us to abandon the most elementary properties of the integral and would therefore be of little interest or value. In fact if we retain the usual operating rules which apply to the symbol $+\infty$ we should have

$$2\delta(x) \;=\; 0 \quad \text{for} \quad \text{each} \quad x \neq 0,$$

and

$$2\delta(0) \ = \ 2 \cdot (+\infty) \ = \ +\infty$$

so that $2\delta = \delta$. Hence, for $a < 0 < b$ we should get by (0.4),

$$\int_a^b 2\delta(x)\,dx \ = \ \int_a^b \delta(x)\,dx \ = \ 1.$$

On the other hand, unless the elementary property of the integral of the product of a function by a constant is not satisfied, we should again have by (0.4),

$$\int_a^b 2\delta(x)\,dx \ = \ 2\int_a^b \delta(x)\,dx \ = \ 2,$$

which contradicts the preceding result. This clearly suggests that there exists no *function* δ which satisfies both the conditions (0.3) and (0.4), under any acceptable theory of integration.

In spite of contradictions such as this, and of similar difficulties which continued to appear, physicists did not hesitate to construct physical theories and entities which appeared to make it necessary to abandon the classical ideas and to accept new *"generalised functions"* such as the *"derivatives"* $\delta', \delta'', \ldots$ of the Dirac δ function. For example, if we consider the system formed by two symmetrical electrical point charges $+q$ and $-q$ placed, respectively, at the points $-h$ and $+h$ of $\mathbb{R}$, then we get a distribution of charge to which, in line with the previous reasoning, it is natural to assign a density function $q\delta(x + h) - q\delta(x - h)$. Suppose now that $h \to 0$ and $q \to +\infty$ in such a way that the product $2qh$ has a finite limit, say p. Heuristically, the limiting charge density would then be

$$\lim_{h \to 0} \left[2qh\, \frac{\delta(x + h) - \delta(x - h)}{2h} \right] \ = \ p\delta'(x).$$

This charge the physicists would call a *"dipole with moment p placed at the origin"*. Similarly the second derivative of δ, which could be heuristically defined as the limit of a system of two dipoles

$$\lim_{h \to 0} \frac{\delta'(x + h) - \delta'(x)}{h} \ = \ \delta''(x)$$

would be interpreted as a *"quadripole"*, etc.

It is clear that all the preceding considerations which, for the sake of simplicity, we have dealt with in the context of the real axis, have their counterpart in spaces of higher dimension. Thus there will appear distributions concentrated on points, lines, surfaces etc., which similarly lead to new *"generalised functions"*, which again display contradictory features when analysed in accordance with classical Analysis. For example, in the case of the space $\mathbb{R}^3 = \mathbb{R}_x \times \mathbb{R}_y \times \mathbb{R}_z$, to a charge q located at the point (a, b, c) there would correspond the density $q\delta(x - a)\delta(y - b)\delta(z - c)$; similarly the symbols $\varphi(x)\delta(y)\delta(z)$ and $\psi(x,y)\delta(z)$ would represent, respectively, a distribution of charge concentrated on the x-axis with linear density $\varphi(x)$, and a distribution concentrated on the plane $z = 0$ with a surface density $\psi(x,y)$. Examples could be given indefinitely, each of which having no meaning in the context of classical Analysis.

Offering as it does a simple and rigorous extension of the classical context, the theory of distributions has not only allowed entirely satisfactory interpretations of several *"generalised functions"* which are important in aplications, but has also revealed in several branches of mathematics, notably in the theory of differential equations, new and unexpected horizons of as yet unpredictable magnitude.

Chapter 1
Axiomatic System for Distributions. Fundamental Operations.

1.1 Notation. Intuitive Approach to Distributions as Generalised Derivatives of Continuous Functions.

In the sequel we will denote by $\mathbb{N}$ the set of all natural numbers (non-negative integers) and by $\mathbb{N}_1$ the set of all positive integers. $\mathbb{R}$ and $\mathbb{C}$ will denote the set of real numbers and the set of all complex numbers, respectively, both sets being provided with their usual structures. If $a, b \in \mathbb{R}$ and $a < b$, then

$$[a,b] \, , \quad]a,b] \, , \quad [a,b[\, , \quad \text{and} \quad]a,b[$$

are the intervals with end points a and b defined respectively by the conditions:

$$a \leq x \leq b, \ \ a < x \leq b, \ \ a \leq x < b \text{ and } a < x < b.$$

All these intervals are *bounded*; $[a,b]$ is a *closed interval* and $]a,b[$ is an *open interval*.

An interval - or, more generally, any subset of $\mathbb{R}$ - is said to be *compact* if it is bounded and closed.

By a natural extension of the language we will call an interval containing a single point $a \in \mathbb{R}$ a *degenerate interval* (the interval $[a,a]$); the empty set may also be considered to be a degenerate interval.

On the other hand, the symbols

$$[a,+\infty[\, , \quad]a,+\infty[\, , \quad]-\infty,a[\, , \quad]-\infty,a]$$

will denote the *unbounded intervals* defined, respectively, by:

$$x \geq a, \ \ x > a, \ \ x < a, \ \ x \leq a,$$

and the symbol $]-\infty,+\infty[$ will denote $\mathbb{R}$ itself.

We remark here that the word *interval* will always be understood to mean a *non-degenerate* interval (bounded or unbounded), unless the contrary is specifically stated.

As already mentioned in the Introduction, the concept of distribution generalises that of function, and the essential purpose of that generalisation is to make it possible to compute derivatives in situations where classical Analysis does not allow this. In particular, any continuous function on an interval $\mathbf{I}$ of the real line, has derivatives of all orders on $\mathbf{I}$ within the context of the theory of distributions, (while from the classical point of view there are continuous functions which are not differentiable at certain points of $\mathbf{I}$ or even at any points of $\mathbf{I}$).

Clearly, the derivatives of continuous functions under this new theory will not generally be functions but will be examples of the new mathematical objects which we call distributions. Similarly the concept of derivative itself is not the usual one but a generalisation of it. Finally, the expression "defined on $\mathbf{I}$" when applied to distributions will have a meaning distinct from the usual one. In contrast to a function, a distribution defined on an interval $\mathbf{I}$ does not in general assume a specific value at each point of $\mathbf{I}$; the concept of the value of a distribution at a point of its domain can only be defined in very particular situations.

The creation of new mathematical objects to allow an operation such as differentiation to be performed under more general conditions is comparable with other generalisations current in various domains of Mathematics. We may mention just one example which is particularly significant, namely the successive extensions of the concept of number. The fundamental objective of these extensions is to allow certain operations (subtraction, division, etc.) to be carried out without restriction, thereby creating new objects; from the set of natural numbers $\mathbb{N}$ one passes to the integers $\mathbb{Z}$, so that it becomes always possible to subtract; and from the integers one goes over to the set of rationals, $\mathbf{Q}$, so as to make division (almost) always possible. The creation of the irrational numbers is designed to guarantee the possibility of passing to the limit in cases where this is not possible in $\mathbf{Q}$, while the the set $\mathbf{C}$ of complex numbers is created to allow the solution of algebraic equations.

After these preliminary remarks let $\mathcal{C}(\mathbf{I})$ denote the set of all complex valued functions[1] defined and continuous on the non degenerate interval $\mathbf{I}$ (if the interval $\mathbf{I}$ is clearly understood we may write $\mathcal{C}$ instead of $\mathcal{C}(\mathbf{I})$ for the sake of simplicity). We

---

[1] As is well known, a complex valued function F defined on the interval $\mathbf{I} \subset \mathbb{R}$ may always be decomposed into a sum $F_1 + iF_2$, where F_1 and F_2 are real valued functions defined on $\mathbf{I}$. The function F is continuous iff F_1 and F_2 are themselves continuous and is n times continuously differentiable (or of class $\mathcal{C}^n$) iff each one of the functions F_1 and F_2 is. If $a, b \in \mathbf{I}$, the integral

$$\int_a^b F(t)dt$$

will exist iff both the integrals

$$\int_a^b F_1(t)dt \quad \text{and} \quad \int_a^b F_2(t)dt,$$

exist, thus verifying the equality

$$\int_a^b F(t)dt \;=\; \int_a^b F_1(t)dt + i \int_a^b F_2(t)dt$$

("iff" is an abreviation of "if and only if").

start by emphasizing that just as it is not always possible to subtract in $\mathbb{N}$ although we can always add (or, as in $\mathbb{Z}$, that it is not always possible to divide although we can always multiply) so in $\mathcal{C}(\mathbf{I})$ it is not always possible to differentiate although antiderivation is always a permissible operation. In fact any arbitrarily chosen function $F \in \mathcal{C}(\mathbf{I}) = \mathcal{C}$ has primitives: to obtain just one such primitive G, it is enough to put

$$G(x) \;=\; \int_a^x F(t)dt \;\;(x \in \mathbf{I}),$$

where a is any fixed point in the interval $\mathbf{I}$.

In the sequel we will denote by $\mathcal{J}_a F$ (or simply by $\mathcal{J} F$ whenever the point $a \in \mathbf{I}$ is fixed once and for all) the function defined for each $F \in \mathcal{C}$ by the above formula, that is, the primitive of F which is zero at the point a. Thus all primitives of F (on $\mathbf{I}$) will be functions of the form $\mathcal{J} F + c$, where c is an arbitrary complex constant. Denoting, as usual, by $\mathcal{C}^1$ (or $\mathcal{C}^1(\mathbf{I})$ when it is convenient to specify the interval) the set of all functions continuously differentiable on $\mathbf{I}$ we have[2]

$$\mathcal{C}^1 \;=\; \{\mathcal{J} F + c : \; F \in \mathcal{C}, \; c \in \mathbf{C}\}$$

where the inclusion $\mathcal{C}^1 \subset \mathcal{C}$ is clearly satisfied.

More generally, define powers of the "operator" $\mathcal{J}$ by setting

$$\mathcal{J}^0 F = F, \;\; \mathcal{J}^{n+1} F = \mathcal{J}\left(\mathcal{J}^n F\right) \;\;\; (F \in \mathcal{C}, \; n \in \mathbb{N})$$

and denote by $\mathcal{P}_n$ the set of all polynomials of degree less than n (restricted to the interval $\mathbf{I}$ and with complex coefficients).[3] The following equality

$$\mathcal{C}^n \;=\; \{\mathcal{J}^n F + P : \; F \in \mathcal{C}, \; P \in \mathcal{P}_n\},$$

holds, where $\mathcal{C}^n$ is the set of all functions n times continuously differentiable on $\mathbf{I}$.

Finally, setting

$$\mathcal{C}^\infty \;=\; \bigcap_{n \in \mathbf{N}} \mathcal{C}^n$$

(where, of course, $\mathcal{C}^0 = \mathcal{C}$) we obtain for every $n \geq 1$:

$$\mathcal{C}^\infty \subset \cdots \subset \mathcal{C}^{n+1} \subset \mathcal{C}^n \subset \cdots \subset \mathcal{C}^1 \subset \mathcal{C}.$$

For each $n \in \mathbb{N}$, the derivation operator may be understood as a mapping (surjective, not injective) from $\mathcal{C}^{n+1}$ into $\mathcal{C}^n$ (or from $\mathcal{C}^\infty$ into itself). More specifically, taking into account the usual algebraic structure of $\mathcal{C}$ as a complex linear space, each space $\mathcal{C}^n$

---

[2] In case $\mathbf{I} = [a,b]$, with $a,b \in \mathbb{R}$ and $a < b$, the function $F : \mathbf{I} \to \mathbf{C}$ is said to be *continuously differentiable* $(F \in \mathcal{C}^1)$ if there exists a continuous function (denoted by F') which for each point x in $]a,b[$ is equal to $F'(x)$ and which at the points a and b takes the values $F_r'(a)$ and $F_l'(b)$, respectively; for $n > 1$, $F \in \mathcal{C}^n$ iff $F' \in \mathcal{C}^{n-1}$. Similar conventions are used when the interval $\mathbf{I}$ contains only one end-point.

[3] $\mathcal{P}_0$ denotes the singular set which comprises the null polynomial.

(with $n \in \mathbb{N}$ or $n = \infty$) will be a linear subspace of $\mathcal{C}$ and the derivation operator will be a linear map from $\mathcal{C}^{n+1}$ into $\mathcal{C}^n$ (or of $\mathcal{C}^\infty$ into itself) of which the kernel is the set $\mathcal{P}_1$ of the constant functions.

Our aim now is to extend the previous chain of spaces in the opposite direction in order to obtain another chain of the form

$$\mathcal{C} \subset \mathcal{C}_1 \subset \ldots \subset \mathcal{C}_n \subset \mathcal{C}_{n+1} \subset \ldots \subset \mathcal{C}_\infty,$$

where $\mathcal{C}_1$ will be regarded, intuitively, as the set of all (generalised) first derivatives of the functions $F \in \mathcal{C}$ and, in general, $\mathcal{C}_n$ will be the set of all such derivatives of order n of the same functions. Finally,

$$\mathcal{C}_\infty = \bigcup_{n \in \mathbb{N}} \mathcal{C}_n$$

will be the set of all distributions.

The rigorous formulation of the preceeding ideas will be the object of the next sections.

1.2   Axiomatic System for Distributions.

In this section four axioms will be stated so as to characterize in a certain way the distributions defined on an interval of $\mathbb{R}$. Where appropriate, some comments will be inserted between consecutive axioms in order to motivate them and to clarify their meaning.

Naturally some fundamental notions from Analysis, such as those of continuous function and derivative (in the classical sense) will be assumed previously known. In particular, before the introduction of the axioms we will assume the meaning of symbols such as $\mathcal{C}$ and $\mathcal{C}^n$ (with $n \in \mathbb{N}$ or $n = \infty$) to be clearly understood, although the same does not of course apply to such symbols as $\mathcal{C}_n$ or $\mathcal{C}_\infty$.

The system of axioms contains two primitive terms: *distribution* and (generalised) *derivative*. To begin with, therefore, the meaning of these two terms is supposed to be completely undetermined; the content will be progressively clarified and made precise as the propositions whose truth must be accepted without proof (that is the axioms of the theory) are stated.

In what follows $\mathbf{I}$ will denote an interval of $\mathbb{R}$ which is subject only to the restriction that it will be non-degenerate; moreover symbols like $\mathcal{C}(\mathbf{I})$, $\mathcal{C}^1(\mathbf{I})$, etc. will be shortened to $\mathcal{C}$, $\mathcal{C}^1$, ... as already mentioned.

> **Axiom 1.** The distributions defined on $\mathbf{I}$ constitute a set, denoted by $\mathcal{C}_\infty(\mathbf{I})$ or $\mathcal{C}_\infty$, which contains the set $\mathcal{C}$ of all continuous functions defined on $\mathbf{I}$.

As expected we have $\mathcal{C} \subset \mathcal{C}_\infty$: every continuous function is a distribution.

Axiom 2. The derivative is a mapping from $\mathcal{C}_\infty$ into $\mathcal{C}_\infty$, denoted by the symbol $\mathbf{D}$, which satisfies the following condition: whenever a distribution f belongs to the set $\mathcal{C}^1$, that is, is a function continuously differentiable on $\mathbf{I}$, then $\mathbf{D}f = f'$ (where f' denotes the function defined and continuous on $\mathbf{I}$ which is the usual classical derivative of the function f).

Thus, for example, (if $\mathbf{I} = \mathbb{R}$), the generalised derivative $\mathbf{D}F$ of the function F such that $F(x) = x|x|$ will be its usual derivative, $F'(x) = 2|x|$. However for the function $\psi \in \mathcal{C}$ defined by $\psi(x) = x^2 \sin \frac{1}{x}$ for $x \neq 0$ which is differentiable on $\mathbb{R}$ but which has a derivative discontinuous at the origin, there is no immediate relation between the distributional derivative $\mathbf{D}\psi$ and the usual derivative ψ', and therefore we cannot (as yet) consider ψ' as a distribution.

It is convenient to remark immediately that defining as usual the power $\mathbf{D}^n$ of $\mathbf{D}$ as follows

$$\mathbf{D}^0 f = f, \quad \mathbf{D}^{n+1}f = \mathbf{D}(\mathbf{D}^n f) \quad (f \in \mathcal{C}_\infty; \; n \in \mathbb{N}),$$

any symbol of the form $\mathbf{D}^n F$, with $F \in \mathcal{C}$, will represent a distribution; this can be immediately deduced from the preceding axioms (and this is what should be verified since the aim of the new theory is precisely that any continuous function on $\mathbf{I}$ should have derivatives of all orders). Thus we may consider as already defined a mapping from $\mathcal{C} \times \mathbb{N}$ into $\mathcal{C}_\infty$ which to each pair $(F, n) \in \mathcal{C} \times \mathbb{N}$ makes correspond the distribution $\mathbf{D}^n F$.

The following axiom makes it clear that this mapping is onto; this, in a way, expresses a certain kind of "economy": there are no more distributions[4] than those which are needed to ensure that there exist derivatives of all orders for all continuous functions.

Axiom 3. For any $f \in \mathcal{C}_\infty$ there exist $F \in \mathcal{C}$ and $n \in \mathbb{N}$ such that $f = \mathbf{D}^n F$.

To clarify the relations between distributions, so called, and symbols of the form $\mathbf{D}^n F$, or even better, the pairs $(F, n) \in \mathcal{C} \times \mathbb{N}$, it remains to be shown when two of these pairs do represent the same distribution. For, it is obvious that the map from $\mathcal{C} \times \mathbb{N}$ into $\mathcal{C}_\infty$ under consideration is not injective. For example, if F is a function which is twice continuously differentiable, then by axioms 1 and 2 the equalities $\mathbf{D}^2 F = \mathbf{D}F' = F''$ must hold and therefore the pairs $(F, 2), (F', 1), (F'', 0)$ must correspond to the same distribution.

It is convenient at this point to recall the corresponding situation when defining the concept of rational number in terms of the given structure of $\mathbb{Z}$. Taking into account that what is intended is to make it possible to divide any integer p by any

---

[4]It is important to remark that in a more general theory which will be described on section 3.4 axiom 3 is not satisfied: in that theory there are distributions - said to be of *infinite order* or of *infinite degree* - which cannot be represented as $\mathbf{D}^n F$, with $F \in \mathcal{C}$ and $n \in \mathbb{N}$.

other integer q (with $q \neq 0$), one is led to conceive each rational number as a quotient $\frac{p}{q}$; however, in general, these quotients make no sense until the rationals themselves have been constructed. It is thus natural to think in terms not of quotients but in terms of ordered pairs $(p, q) \in \mathbb{Z} \times (\mathbb{Z} \backslash \{0\})$, which already do have a precise sense. It is natural to demand that each one of these pairs should determine a rational number, but here also two distinct pairs may determine the same element in $\mathbf{Q}$ (for example, the rationals corresponding to the pairs $(1, 3)$ and $(2, 6)$ - or, if one prefers, to the "quotients" $\frac{1}{3}$ and $\frac{2}{6}$ - must be identified in some sense).

What condition, then, must two pairs (p, q) and (r, s) satisfy so that they correspond to the same rational number? Such a condition cannot be expressed directly in the form $\frac{p}{q} = \frac{r}{s}$, which has no meaning, in general, within the set $\mathbb{Z}$; but it may be expressed through the equality $ps = qr$. It is easy to verify that this equality determines an equivalence relation on the set $\mathbb{Z} \times (\mathbb{Z} \backslash \{0\})$, and that to each rational number there may be associated a corresponding equivalence class; hence, the set $\mathbf{Q}$ may be interpreted as the quotient of the set $\mathbb{Z} \times (\mathbb{Z} \backslash \{0\})$ with respect to that equivalence relation.

However, it is clear that in practice when rational numbers are used nobody thinks of equivalence classes: what is important then is not the nature of these numbers but the formal rules to operate with them, the "rules of the game", which are expressed in appropriate definitions. In the same way, in the case of the axiomatics of distributions, it is necessary to fix the relation that two pairs $(F, m), (G, n) \in \mathcal{C} \times \mathbb{N}$ must satisfy so that they determine the same distribution. That relation cannot be expressed by the equality $\mathbf{D}^m F = \mathbf{D}^n G$, which has no sense, in general, until the set of distributions has been constructed. However, instead of thinking in terms of derivatives of the continuous functions F and G, it is natural at this stage to try to obtain the desired result by making appeal to the operation of antiderivation, which is always possible in $\mathcal{C}$.

To clarify ideas it is convenient to start with the simplest cases: for example, we must consider when does the equality $\mathbf{D} F = \mathbf{D} G$ hold or, equivalently, under what conditions should the pairs $(F, 1)$ and $(G, 1)$ be identified? In the particular case when F and G both belong to $\mathcal{C}^1$ the answer is immediate. In fact, under that hypothesis, by the axiom 2, we will have that $\mathbf{D} F = F'$ and $\mathbf{D} G = G'$ (where F' and G' are the usual derivatives of F and G, respectively) and therefore the equality $\mathbf{D} F = \mathbf{D} G$ must be equivalent to $(F - G)' = 0$. This will be satisfied only when the function $F - G$ is a constant function, $(F - G) \in \mathcal{P}_1$.

Nothing is then more natural than to accept this condition as the requirement for the equality $\mathbf{D} F = \mathbf{D} G$ to hold even in the case when the functions F and G are simply arbitrary elements in $\mathcal{C}$, not necessarily belonging to $\mathcal{C}^1$. More generally, it is easy to see that if $r \in \mathbb{N}$ and $F, G \in \mathcal{C}^r$, then the equality $\mathbf{D}^r F = \mathbf{D}^r G$ is satisfied iff $F - G \in \mathcal{P}_r$. It therefore becomes natural to adopt as the last axiom for the theory, the following

Axiom 4. If $F, G \in \mathcal{C}$ and $r \in \mathbb{N}$ then the equality $\mathbf{D}^r F = \mathbf{D}^r G$ holds iff $F - G \in \mathcal{P}_r$.

It is now easy to obtain conditions under which the equality $\mathbf{D}^m F = \mathbf{D}^n G$, with $m, n \in \mathbb{N}$ and $F, G \in \mathcal{C}$, holds. For that purpose let us start by observing that whenever $F \in \mathcal{C}$ (and therefore $\mathcal{J} F \in \mathcal{C}^1$) one must have $\mathbf{D}(\mathcal{J} F) = (\mathcal{J} F)' = F$; more generally, for every natural number p under the same hypothesis we must have $\mathbf{D}^p(\mathcal{J}^p F) = F$.

From these considerations there arises the possibility of representing two arbitrary distributions, $\mathbf{D}^m F$ and $\mathbf{D}^n G$ as derivatives of continuous functions of the same order; in fact it is enough to see that

$$\mathbf{D}^m F \;=\; \mathbf{D}^m \left(\mathbf{D}^n \mathcal{J}^n F\right) \;=\; \mathbf{D}^{m+n} \left(\mathcal{J}^n F\right),$$

$$\mathbf{D}^n G \;=\; \mathbf{D}^n \left(\mathbf{D}^m \mathcal{J}^m G\right) \;=\; \mathbf{D}^{m+n} \left(\mathcal{J}^m G\right).$$

From axiom 4 it then follows that the equality $\mathbf{D}^m F = \mathbf{D}^n G$ holds if and only if $\mathcal{J}^n F - \mathcal{J}^m G \in \mathcal{P}_{m+n}$.

1.3 Consistency and Categoricity of the System of Axioms.

Once the formal system of axioms is set up the following question arises: can we in fact admit the existence of distributions (that is, do there really exist mathematical objects satisfying axioms 1 through 4) without fear of introducing any contradiction in the mathematical theory that we are constructing?

To prove that the answer to this question is in the afirmative, that is, to prove the *compatibility* of the axiomatic system, it is convenient to construct a *model*, by defining a set (whose elements we will call *distributions*) and a mapping from that set into itself (which we will call *derivation*) such that as a result, the four axioms under consideration are satisfied.

Accordingly we begin by proving that the relation ρ, defined on $\mathcal{C} \times \mathbb{N}$ by the formula

$$(F, m) \, \rho \, (G, n) \;\Leftrightarrow\; \mathcal{J}^n F - \mathcal{J}^m G \in \mathcal{P}_{m+n}$$

is an equivalence relation. Since reflexivity and symmetry are obvious, we only require to prove the transitivity.

Supposing that $(F, m) \, \rho \, (G, n)$ and $(G, n) \, \rho \, (H, p)$, that is, supposing that

$$\mathcal{J}^n F - \mathcal{J}^m G \in \mathcal{P}_{m+n} \quad \text{and} \quad \mathcal{J}^p G - \mathcal{J}^n H \in \mathcal{P}_{n+p},$$

and taking into account that the antiderivatives of order k of a polynomial of degree $< r$ are polynomials of degree less than $r + k$, we have

$$\mathcal{J}^{n+p} F - \mathcal{J}^{m+p} G \in \mathcal{P}_{m+n+p} \,, \quad \mathcal{J}^{m+p} G - \mathcal{J}^{m+n} H \in \mathcal{P}_{m+n+p}.$$

From these relations by subtraction we get

$$\mathcal{J}^{n+p}F - \mathcal{J}^{n+m}H \in \mathcal{P}_{m+n+p}$$

or, taking the nth derivative,

$$\mathcal{J}^{p}F - \mathcal{J}^{m}H \in \mathcal{P}_{m+p}$$

that is $(F,m)\,\rho\,(H,p)$.

Hence ρ is an equivalence relation.

For each pair $(F,m) \in \mathcal{C} \times \mathbb{N}$ let us denote by $[F,m]$ the ρ-equivalence class to which the pair (F,m) belongs; let us also denote by $\mathcal{C}_\infty^*$ the set of all such equivalence classes, that is, the quotient of $\mathcal{C} \times \mathbb{N}$ with respect to the relation ρ:

$$\mathcal{C}_\infty^* \; = \; \mathcal{C} \times \mathbb{N}/\rho \; = \; \{[F,m] : F \in \mathcal{C},\; m \in \mathbb{N}\},$$

where

$$[F,m] \; = \; \{(G,n) \in \mathcal{C} \times \mathbb{N} : (G,n)\rho(F,m)\}.$$

On the other hand, for each $F \in \mathcal{C}$, we identify[5] the function F with the equivalence class $[F,0]$. We now verify that, by the formula

$$\mathbf{D}_*\,([F,m]) \; = \; [F,m+1],$$

we define a map $\mathbf{D}_*$ from $\mathcal{C}_\infty^*$ into itself; it is enough to notice that if we have $(F,m)\,\rho\,(G,n)$ then the relation $(F,m+1)\,\rho\,(G,n+1)$ also holds, that is, from $\mathcal{J}^{n}F - \mathcal{J}^{m}G \in \mathcal{P}_{m+n}$ it follows immediately that $\mathcal{J}^{n+1}F - \mathcal{J}^{m+1}G \in \mathcal{P}_{m+n+2}$ (we can even prove that $\mathcal{J}^{n+1}F - \mathcal{J}^{m+1}G \in \mathcal{P}_{m+n+1}$).

It is now easy to see that, if in the statements of the axioms 1 to 4 we replace $\mathcal{C}_\infty$ by $\mathcal{C}_\infty^*$ and $\mathbf{D}$ by $\mathbf{D}_*$ we will always obtain true propositions.

For axiom 1 this follows from the identification of $\mathcal{C}$ with that part of $\mathcal{C}_\infty^*$ referred to above. To complete the verification of axiom 2 it is enough to note that if an element in $\mathcal{C}_\infty^*$ is identified with a function $F \in \mathcal{C}^1$ that element is $[F,0]$ and, therefore, taking into account that

$$\mathbf{D}_*\,([F,0]) \; = \; [F,1] \; = \; [F',0],$$

its derivative (in the sense of $\mathbf{D}_*$) is identical to F', that is, to its usual derivative.

Axiom 3 is also immediately verifiable: an arbitrary element $[F,m]$ in $\mathcal{C}_\infty^*$ may always be represented in the form

$$\mathbf{D}_*^{m}\,([F,0]) \; = \; \mathbf{D}_*^{m}F.$$

---

[5]To identify two mathematical objects is to accept that both may be designated by the same symbol. Note that it is similarly necessary to make such an identification in order that each integer number may also be considered as a rational number (it is then an identification of an integer number p with the equivalence class which contains the pair $(p,1) \in \mathbb{Z} \times \mathbb{Z}\backslash\{0\}$).

Finally if we have $\mathbf{D}_*^r F = \mathbf{D}_*^r G$, that is to say if $[F, r] = [G, r]$, then we also have that $\mathcal{J}^r F - \mathcal{J}^r G \in \mathcal{P}_{2r}$ and, therefore, $F - G \in \mathcal{P}_r$ which confirms axiom 4.

Thus we have obtained a model of the axiomatic system, and this is in a sense equivalent to a proof of the existence of distributions.

Another important question refers to the "uniqueness" of the distributions defined on a given interval $\mathbf{I}$ or, in more rigorous terms, to the *categoricity* of the axiomatic system. Clearly the reasonable way to formulate this question is not to ask if the model obtained is unique[6] but instead to ask if any other model will necessarily be isomorphic to this one. In other words: if $\mathcal{C}_\infty^0$ is a set and $\mathbf{D}_0$ a map of $\mathcal{C}_\infty^0$ into itself such that the replacement of $\mathcal{C}_\infty$ and $\mathbf{D}$ respectively by $\mathcal{C}_\infty^0$ and $\mathbf{D}_0$ transforms the axioms 1 through 4 into true propositions, then can we guarantee that there exists a bijective mapping $\varphi : \mathcal{C}_\infty^0 \to \mathcal{C}_\infty^*$ which respects derivation (that is, such that $\varphi(\mathbf{D}_0 f) = \mathbf{D}_* \varphi(f)$ for each $f \in \mathcal{C}_\infty^0$)?

As will be seen in the sequel the answer to this question is in the affirmative, and it is even possible to define an isomorphism φ leaving invariant the continuous functions (that is, an isomorphism φ such that $\varphi(F) = F$, whenever $F \in \mathcal{C}$). In fact having once defined such a map φ on the subset $\mathcal{C}$ of $\mathcal{C}_\infty^0$ we can extend it to the whole of $\mathcal{C}_\infty^0$ as follows: for each distribution $f \in \mathcal{C}_\infty^0$ (which, according to axiom 3, is of the form $f = \mathbf{D}_0^n F$ with $n \in \mathbb{N}$ and $F \in \mathcal{C}$), set

$$\varphi(f) \; = \; \varphi(\mathbf{D}_0^n F) \; = \; \mathbf{D}_*^n \varphi(F) \; = \; \mathbf{D}_*^n F.$$

That this really does define a mapping from $\mathcal{C}_\infty^0$ into $\mathcal{C}_\infty^*$ is a consequence of the fact that the criterion of equality for the generalised derivatives of continuous functions is the same in both cases (this criterion following immediately from the axioms, which both structures are supposed to satisfy); hence, if $\mathbf{D}_0^m F = \mathbf{D}_0^n G$ then $\mathcal{J}^n F - \mathcal{J}^m G$ belongs to $\mathcal{P}_{m+n}$ and therefore $\mathbf{D}_*^m F = \mathbf{D}_*^n G$.

The mapping $\varphi : \mathcal{C}_\infty^0 \to \mathcal{C}_\infty^*$ is a surjective map in view of axioms 2 and 3. It is injective since if $f = \mathbf{D}_0^m F, g = \mathbf{D}_0^n G$ and $\varphi(f) = \varphi(g)$, that is, $\mathbf{D}_*^m F = \mathbf{D}_*^n G$, then we also have $\mathcal{J}^n F - \mathcal{J}^m G \in \mathcal{P}_{m+n}$ and therefore $f = g$. Finally for any distribution $f = \mathbf{D}_0^m F \in \mathcal{C}_\infty^0$ we have

$$\varphi(\mathbf{D}_0 f) \; = \; \varphi\left(\mathbf{D}_0^{m+1} F\right) \; = \; \mathbf{D}_*^{m+1} F \; = \; \mathbf{D}_*\left(\mathbf{D}_*^m F\right) \; = \; \mathbf{D}_* \varphi(f).$$

We can therefore say that the uniqueness of the set of distributions over an interval $\mathbf{I}$ and of the corresponding generalised derivation operator is now, in a certain sense, ensured.

---

[6] In this very restricted sense it is obvious that such "uniqueness" would not hold: it would be enough, for example, to repeat all steps of the construction of the above model considering, instead of pairs of the form $(F, m) \in \mathcal{C} \times \mathbb{N}$, pairs of the form $(m, F) \in \mathbb{N} \times \mathcal{C}$ with all the obvious modifications, to obtain a model which is distinct from the previous one.

1.4   First Examples of Distributions.

Before going any further, it seems convenient to present some concrete examples of distributions.

Naturally, the first examples of distributions are the continuous functions themselves; however, in addition to these, there are other functions which can clearly be interpreted as distributions. We mention in the first place those discontinuous functions which have a continuous primitive. An example which we have already mentioned is the function G such that

$$G(x) = \begin{cases} 2x\sin(\frac{1}{x}) - \cos(\frac{1}{x}) & \text{if} \quad x \neq 0 \\ 0 & \text{if} \quad x = 0 \end{cases}$$

which is discontinuous at the origin, but which has for a primitive the function ψ defined by

$$\psi(x) = \begin{cases} x^2\sin(\frac{1}{x}) & \text{if} \quad x \neq 0 \\ 0 & \text{if} \quad x = 0. \end{cases}$$

It is then natural to identify G with the generalised derivative of the continuous function ψ and with this sense to write $G = \mathbf{D}\psi$.

Another important example of functions which may be identified with distributions, in a sense similar to the previous one, is provided by the class of those functions which on a given interval have an indefinite integral. If $\mathbf{I}$ is an interval of $\mathbb{R}$ a function F is said *to have an indefinite integral* over $\mathbf{I}$ iff F is Riemann integrable[7] over every compact subinterval of $\mathbf{I}$. If F is such a function and a is an arbitrarily fixed point in $\mathbf{I}$ the *indefinite integral of F with origin at the point a* is the function continuous on $\mathbf{I}$ which is defined by

$$\Phi(x) \; = \; \int_a^x F(t)dt.$$

Usually the function F will be defined everywhere on the interval $\mathbf{I}$ (or on any other set containing that interval); for the sake of generality, however, it is convenient to accept the possibility that F may not be defined on some "exceptional" points of $\mathbf{I}$, but is always such that it is possible to give a precise sense to the indefinite integral.[8]

It is well-known that at any point $x \in \mathbf{I}$ of continuity of F, the indefinite integral Φ of F will be differentiable and that in this case we will have $\Phi'(x) = F(x)$. In particular if $F \in \mathcal{C}(\mathbf{I})$ we will have that $\Phi \in \mathcal{C}^1(\mathbf{I})$ and $\Phi' = F$. It then becomes natural, in the more general case when F is any function with indefinite integral, to

---

[7]Any reader familiar with the theory of the Lebesgue integral may adopt this sense of integrability here with advantage; the concept of "function with indefinite integral" must in this case be replaced by "locally integrable function".

[8]That is to say, it must be accepted that : i) it is possible to attribute values to the function F at the "exceptional" points so that we can obtain a function which is integrable between a and x for any $a, x \in \mathbf{I}$; ii) no other values at the same points lead to a function which also has an indefinite integral over $\mathbf{I}$ for which the integral between a and x has a different value (for some points $a, x \in \mathbf{I}$).

identify F with the distributional derivative of its indefinite integral, and with this sense to write $F = \mathbf{D}\Phi$.

Consider, for example, the case of the function (with indefinite integral on $\mathbb{R}$):

$$\mathbf{H}(x) = \begin{cases} 1 & \text{if} \quad x > 0 \\ 0 & \text{if} \quad x < 0. \end{cases}$$

Choosing for example $a = 0$ and setting

$$J(x) \;=\; \int_0^x \mathbf{H}(t)dt$$

(that is, $J(x) = x$ if $x \geq 0$ and $J(x) = 0$ if $x < 0$) the process just described leads to the identification of the function $\mathbf{H}$, called the *Heaviside function*, with the distribution $\mathbf{D}J$, which will be described as the *Heaviside distribution.*

It is convenient to observe, however, that besides this Heaviside function there are infinitely many other functions which, in this way may also be identified with this same distribution. For instance any function which coincides with $\mathbf{H}$ at all points of the set $\mathbb{R}\backslash\mathbb{Z}$ will also be identified with the distribution $\mathbf{D}J$ independently of the values which it assumes at the points of $\mathbb{Z}$ (and even if it is not defined at all on such points).

More generally, if F and G are two functions with indefinite integrals on the interval $\mathbf{I}$ and if, given a (fixed) point $a \in \mathbf{I}$, we have

$$\mathcal{J}_a F(x) \;=\; \int_a^x F(t)dt \;=\; \int_a^x G(t)dt \;=\; \mathcal{J}_a G(x)$$

at every point $x \in \mathbf{I}$, then we also have $\mathbf{D}\left(\mathcal{J}_a F\right) = \mathbf{D}\left(\mathcal{J}_a G\right)$ and therefore the two functions will be identified with one and the same distribution. Conversely, if there exists any $x \in \mathbf{I}$ such that $\mathcal{J}_a F(x) \neq \mathcal{J}_a G(x)$, the difference $\mathcal{J}_a F(x) - \mathcal{J}_a G(x)$ (which is zero at the point a) will not be constant on $\mathbf{I}$; and, as is an easy consequence of the axioms, since the only distributions with null derivative are the constants, we can conclude that $\mathbf{D}\left(\mathcal{J}_a F\right) \neq \mathbf{D}\left(\mathcal{J}_a G\right)$; that is, F and G will be identified with distinct distributions.

Of course we shall say that the functions F and G *have the same indefinite integral on* $\mathbf{I}$ iff for any $a, x \in \mathbf{I}$, the integrals[9]

$$\int_a^x F(t)dt \quad \text{and} \quad \int_a^x G(t)dt$$

---

[9]Given the existence of the integrals, for their equality it can be proved that it is necessary and sufficient that we have $F(x) = G(x)$ at every point $x \in \mathbf{I}$ where both functions F and G are continuous; or that, for every compact $\mathbf{K} \subset \mathbf{I}$ we have

$$\int_{\mathbf{K}} |F(t) - G(t)|dt \;=\; 0.$$

both exist and are equal.

The preceding considerations show that the process of identification to which we have been referring does not allow us rigorously to consider the set of functions which have indefinite integral as a part of the set of distributions on **I**. This is so since the mapping from the first set to the second is not an injective map. Only if we first identify all functions which have the same indefinite integral[10] can we say that any function with indefinite integral is a distribution.

Once such identification is accepted every one of those functions will have a derivative in the generalised sense and even derivatives of all orders (since the same is true for distributions). But it may now happen that the usual derivative of such a function F (defined at all points where F is differentiable) will be identified with a distribution which is distinct from the distributional derivative of F. For example the Heaviside function, **H**, which we have identified with the distribution **D**J, has in the usual sense a derivative **H**$'$ defined on $\mathbb{R}\backslash\{0\}$ ($\mathbf{H}'(x) = 0$ for every $x \neq 0$) which, by the process considered above is identified with the null distribution (that is, with the function which is zero at all points of $\mathbb{R}$). However, the derivative of $\mathbf{H} = \mathbf{D}J$ in the generalised sense is the so-called Dirac distribution, $\mathbf{D}^2J$, which is usually denoted by the symbol δ. Moreover, it is clear that since **D**J is not constant over $\mathbb{R}$, δ is not the null distribution.[11] Other examples of distributions which, like δ, are also not identifiable with functions are the derivatives of δ, generally denoted by δ', δ'', ..., $\delta^{(n)}$, ... ($\delta' = \mathbf{D}^3J$, ..., $\delta^{(n)} = \mathbf{D}^{n+2}J$).

Before giving some more examples some remarks about notation are in order. They apply not only in this chapter but also in the sequel.

As is well known, it is common in the mathematical literature to use symbols such as $F(x)$, $F(t)$, ..., to denote the function that should properly be denoted by F. This is obviously an abuse of notation: in fact, if x is a point of the domain of the function F, $F(x)$ will denote the value that F assumes at the point x and not the function itself. In some areas of Mathematics such abuse may cause confusion, sometimes of a serious nature. Accordingly, whenever there is some advantage to be gained in so doing, it is sometimes recommended to use a convention due to Russell in which the symbols $F(\hat{x})$, $F(\hat{t})$, etc. are used instead of F, while the symbol $F(x)$ is reserved to denote the value of F at the point x of its domain. Thus the hat over the variable indicates that it is a dummy variable. Sometimes we shall use this convention but, when there is no danger of confusion, we shall also allow the abuse of notation involved in writing $F(x)$ instead of F; and we will write $f(\hat{x})$, or simply $f(x)$, to denote distributions (although, as already mentioned, in the case of distributions the symbol $f(x)$ does not even denote, in general, the value of f at the point x of its domain, since this is a

---

[10]More precisely, if we replace the set of all functions with indefinite integral by the quotient of this set with respect to the equivalence relation ρ defined by "$F\rho G$ iff F and G have the same indefinite integral on **I**".

[11]Clearly the distribution δ is the counterpart, in rigorous terms, of the Dirac "function" already mentioned in the Introduction.

concept whose meaning is defined only for some particular situations).

Coming back to the examples (and committing the abuse of notation just mentioned), consider now the function $1/\sqrt{|x|}$ which is not bounded and therefore not integrable in the proper Riemann sense over any interval containing the origin. This function does not have an indefinite integral over the interval $\mathbf{I} = \mathbb{R}$ in the sense described above. However, for every $x \in \mathbb{R}$, the improper integral

$$\int_0^x \frac{1}{\sqrt{|t|}}\, dt$$

is convergent, thus defining a function θ which is continuous on $\mathbb{R}$ and such that $\theta'(x) = 1/\sqrt{|x|}$ at every point $x \neq 0$ (that is, at each point of continuity of the function $1/\sqrt{|x|}$). In this case therefore it still seems natural to identify the function considered with the distribution $\mathbf{D}\theta$.

Similarly we can interpret as a distribution on $\mathbb{R}$ the function $\log|x|$ for which we have

$$\int_0^x \log|t|\, dt \;=\; \lambda(x)$$

where λ is that member of $\mathcal{C}(\mathbb{R})$ such that $\lambda(x) = x(\log|x| - 1)$ for every $x \neq 0$; the function $\log|x|$ will therefore be identified with the distribution $\mathbf{D}\lambda$.

We remark that nevertheless there are functions which cannot be identified with a given distribution following this process: this is the case for the function $1/x$, which cannot be interpreted as a distribution on $\mathbb{R}$ (although it can be interpreted as an element in $\mathcal{C}_\infty(\mathbf{I})$ for every interval $\mathbf{I}$ which does not contains the origin). In fact it is immediately recognizable that there does not exist any function Φ continuous on $\mathbb{R}$ such that the condition $\Phi'(x) = 1/x$ holds for all $x \neq 0$. It is enough to note that the general expression for all antiderivatives of $1/x$ on $\mathbb{R}\backslash\{0\}$ is

$$\log|x| + a + b\mathbf{H}(x),$$

where $a, b \in \mathbf{C}$; none of these functions has a finite limit when $x \to 0$.

We might try to consider instead of "antiderivatives of first order", a function $\psi \in \mathcal{C}(\mathbb{R})$ such that $\psi''(x) = 1/x$ for $x \neq 0$, in order to identify the function $1/x$ with the distribution $\mathbf{D}^2\psi$. However it is immediately verifiable that the general expression of such "antiderivatives of second order" is

$$\psi(x) \;=\; \lambda(x) + ax + b + cJ(x)$$

where $a, b, c \in \mathbf{C}$ and $\lambda(x) = x\log|x| - x$ with $\lambda(0) = 0$. Hence, since there is no reason to choose any one of these antiderivatives, one would be led not to a unique distribution but to the set of all distributions of the form $\mathbf{D}^2\lambda + c\delta$, with c a constant.

We remark that this fact may have been the origin of a paradoxical equality

$$\frac{1}{x} \;=\; \frac{1}{x} + c\,\delta$$

which has been given by Dirac in his book on Quantum Mechanics. In fact $1/x$ must not be interpreted as a distribution, but rather as an infinite class of distributions any two of which differ by a constant multiple of δ. Among all these infinitely many distributions which correspond in this way to the function $1/x$ it is sometimes recommended to distinguish one particular distribution, usually described as "the finite part of x^{-1}", $\mathbf{Fp}\{x^{-1}\}$. It is the generalised derivative of $\log|x|$

$$\mathbf{Fp}\{x^{-1}\} \; = \; \mathbf{D}\log|x|$$

to which we will refer again in the sequel.

More generally, for any $n \in \mathbb{N}_1$ the "finite part of x^{-n}" is defined by

$$\mathbf{Fp}\{x^{-n}\} \; = \; \mathbf{D}^n\left[(-1)^{n-1}\frac{1}{(n-1)!}\,\log|x|\right] \, ,$$

it being easily verified (after we give some definitions and state some elementary properties in the next section) that we have, for each $n \in \mathbb{N}_1$:

$$\mathbf{D}\left(\mathbf{Fp}\left\{\frac{1}{x^n}\right\}\right) \; = \; -n\,\mathbf{Fp}\left\{\frac{1}{x^{n+1}}\right\}.$$

1.5   Degree of a Distribution. Addition and Multiplication by a Scalar. Primitives.

For every $n \in \mathbb{N}$ denote by $\mathcal{C}_n(\mathbf{I})$ (or simply $\mathcal{C}_n$) the set of all distributions $f \in \mathcal{C}_\infty(\mathbf{I})$ which satisfy the following condition: there exists a function $F \in \mathcal{C}(\mathbf{I})$ such that $f = \mathbf{D}^n F$. The relation $\mathbf{D}^n F = \mathbf{D}^{n+1}(\mathcal{J}_a F)$, which is valid for every $F \in \mathcal{C}$, $n \in \mathbb{N}$ and $a \in \mathbf{I}$, clearly shows that we always have $\mathcal{C}_n \subset \mathcal{C}_{n+1}$. On the other hand from axiom 3 it follows that

$$\bigcup_{n\in\mathbb{N}} \mathcal{C}_n \; = \; \mathcal{C}_\infty.$$

The *degree* of a distribution f will be defined to be the smallest integer $n \geq 0$ such that $f \in \mathcal{C}_n$; $\mathcal{C}_n$ is therefore the set of all distributions of degree $\leq n$.

The distributions of degree 0 are the continuous functions. On the other hand, we have:

> **Theorem 1.1** *If the distribution f is not a function of class $\mathcal{C}^1$ and is of degree n, then $\mathbf{D}f$ is of degree $n+1$.*

Proof: In case $n = 0$, f is a continuous function and, since it does not belong to $\mathcal{C}^1$, it is clear that $\mathbf{D}f$ is of degree 1. Suppose now that $n > 0$ and that f is of degree n and therefore is representable in the form $f = \mathbf{D}^n F$ with $F \in \mathcal{C}$. Then $\mathbf{D}f = \mathbf{D}^{n+1}F$ which shows that $\mathbf{D}f$ is of degree less than or equal to $n+1$. If it were less than $n+1$ then an equality of the form $\mathbf{D}f = \mathbf{D}^n G$ with $G \in \mathcal{C}$ would hold and given a (fixed) point $a \in \mathbf{I}$, we would have

$$\mathbf{D}^n G \; = \; \mathbf{D}^{n+1}(\mathcal{J}_a G) \; = \; \mathbf{D}^{n+1}F$$

from which by axiom 4

$$F - \mathcal{J}_a G \in \mathcal{P}_{n+1}.$$

However in this case the function F (sum of a function of the class $\mathcal{C}^1$ and a polynomial) would be of class $\mathcal{C}^1$ and therefore $f = \mathbf{D}^n F = \mathbf{D}^{n-1} F'$ would have a degree less than n. $\square$

Thus, since the distribution of Heaviside is of degree 1, it follows that δ is of degree 2, δ' is of degree 3 and, generally, $\delta^{(n)}$ is of degree $n + 2$. From this in particular it follows that for $m \neq n$ we have $\delta^{(m)} \neq \delta^{(n)}$.

Given two arbitrary distributions f and g (defined on the same interval $\mathbf{I}$) it follows immediately from the relation $\mathcal{C}_p \subset \mathcal{C}_{p+1}$ (for every $p \in \mathbb{N}$) that there exists a natural number $n \in \mathbb{N}$ such that f and g both belong to the same set $\mathcal{C}_n$. Then we will have $f = \mathbf{D}^n F$ and $g = \mathbf{D}^n G$ with $F, G \in \mathcal{C}$ and it is very easy to see that the distribution $\mathbf{D}^n(F + G)$ depends only upon f and g (and not on n, F and G, that is, not on the particular representations chosen for the distributions f and g). In fact, for any $m \geq n$, if F^* and G^* are continuous functions such that $f = \mathbf{D}^m F^*$ and $g = \mathbf{D}^m G^*$ we will have

$$\mathbf{D}^m F^* = \mathbf{D}^m \left(\mathcal{J}^{m-n} F \right) , \quad \mathbf{D}^m G^* = \mathbf{D}^m \left(\mathcal{J}^{m-n} G \right)$$

and therefore the functions $P = F^* - \mathcal{J}^{m-n} F$ and $Q = G^* - \mathcal{J}^{m-n} G$ will be polynomials of degree smaller than m. From this fact it follows immediately that

$$\begin{aligned}
\mathbf{D}^m \left(F^* + G^* \right) &= \mathbf{D}^m \left(\mathcal{J}^{m-n} F + P + \mathcal{J}^{m-n} G + Q \right) \\
&= \mathbf{D}^m \mathcal{J}^{m-n}(F + G) = \mathbf{D}^n(F + G),
\end{aligned}$$

which gives the required proof.

The process just described allows us to associate with each pair (f, g) in $\mathcal{C}_\infty \times \mathcal{C}_\infty$ a new distribution thereby defining a binary operation on $\mathcal{C}_\infty$. This operation is naturally called *addition*, and the distribution thus associated with the pair (f, g) is by definition the *sum*, $f + g$, of f and g.

It is an easy matter to see that addition is the unique operation on $\mathcal{C}_\infty$ which enjoys the following properties:

$\mathbf{a_1}$) *if $f, g \in \mathcal{C}$, $f + g$ is the usual sum of the functions f and g;*

$\mathbf{a_2}$) *for any two distributions f and g, $\mathbf{D}(f + g) = \mathbf{D}f + \mathbf{D}g$.*

It is also easy to verify that addition has a zero element, (the null distribution, which will be denoted by $\mathbf{0}$, when no confusion is likely to arise), and that every distribution has an additive inverse: for every $f \in \mathcal{C}_\infty$ there exists $g \in \mathcal{C}_\infty$ such that $f + g = \mathbf{0}$. Moreover we can also see that the properties

$$(f + g) + h = f + (g + h) \text{ and } f + g = g + f$$

hold for any distributions f, g and h. Hence $\mathcal{C}_\infty$ equipped with addition is an *abelian group.*

Similarly we can easily define the product of a distribution f by a complex number α: if $f = \mathbf{D}^n F$ we naturally set $\alpha f = \mathbf{D}^n(\alpha F)$, it being trivial to verify the coherence of this definition (that is, that although the product αf clearly depends on f and α it does not depend on the representation chosen for f).

It is also clear that the multiplication of a distribution by a complex number - that is, the mapping $(\alpha, f) \rightsquigarrow \alpha f$ just defined - is the unique mapping from $\mathbf{C} \times \mathcal{C}_\infty$ into $\mathcal{C}_\infty$ which satisfies the following conditions:

$\mathbf{m_1})$ *if $f \in \mathcal{C}$, αf is the usual product of the number α by the function f;*

$\mathbf{m_2})$ *for every $\alpha \in \mathbf{C}$ and any $f \in \mathcal{C}_\infty$, $\mathbf{D}(\alpha f) = \alpha \mathbf{D} f$.*

It is immediate that if $\alpha, \beta \in \mathbf{C}$ and $f, g \in \mathcal{C}_\infty$ we have

i. $\alpha(f + g) = \alpha f + \alpha g$;

ii. $(\alpha + \beta)f = \alpha f + \beta f$;

iii. $\alpha(\beta f) = (\alpha\beta)f$;

iv. $1f = f$.

From these properties and the fact that $\mathcal{C}_\infty$ is an additive abelian group it follows that $\mathcal{C}_\infty$ *is a linear space over the field* $\mathbf{C}$, or a *complex linear space*. Furthermore since for $n \in \mathbb{N}$ we have $f + g \in \mathcal{C}_n$ and $\alpha f \in \mathcal{C}_n$ whenever $f, g \in \mathcal{C}_n$ and $\alpha \in \mathbf{C}$ then it can be said that every subset $\mathcal{C}_n$ is a linear subspace of $\mathcal{C}_\infty$ (and of $\mathcal{C}_p$ for every $p \geq n$).

It is also very easy to prove that for any two distributions f and g the degree of $f + g$ does not exceed the maximum of the degrees of f and g (being precisely equal to that maximum when the degrees of f and g are unequal). For example, any linear combination of derivatives of the δ distribution, that is, any distribution which may be represented by an expression of the form

$$\sum_{j=0}^{m} a_j \delta^{(j)},$$

is of degree $p + 2$ where p is the greatest index j such that $a_j \neq 0$. If all coefficients a_j are zero, then clearly the degree is zero. From this it follows immediately that an equality of the form

$$\sum_{j=0}^{m} a_j \delta^{(j)} = 0$$

(with $m \in \mathbb{N}$ and $a_j \in \mathbf{C}$, $j = 0, 1, \ldots, m$) holds only if all coefficients are zero.

We should point out that the fact that the properties $\mathbf{a_2})$ and $\mathbf{m_2})$ hold, that is, that for any $f, g \in \mathcal{C}_\infty$ and $\alpha \in \mathbf{C}$ we have

$$\mathbf{D}(f + g) = \mathbf{D}f + \mathbf{D}g$$

$$\mathbf{D}(\alpha f) = \alpha \mathbf{D}f,$$

may be interpreted as saying that the derivation $\mathbf{D}$ is a *linear mapping* of $\mathcal{C}_\infty$ into itself (or from $\mathcal{C}_n$ into $\mathcal{C}_{n+1}$, for $n \in \mathbb{N}$).

We will finish this section with a reference to the antiderivation of distributions. The first fact to remark is that, like derivation itself, antiderivation is an operation always possible in the space of distributions. In fact given a distribution $f = \mathbf{D}^n F$ with $F \in \mathcal{C}$ it is obvious that the distribution $g = \mathbf{D}^n \mathcal{J}_a F$ satisfies the condition $\mathbf{D}g = F$, so that it is a *primitive* of f. It is also easy to see that two primitives of one and the same distribution necessarily differ only by a constant (or, what is equivalent, that the only distributions which have a null derivative are the constants). In fact, if $\mathbf{D}f = 0$ then, representing f by $\mathbf{D}^n F$ with $F \in \mathcal{C}$, we will have $\mathbf{D}^{n+1} F = 0$. By axiom 4, F will therefore be a polynomial of degree $< n + 1$ and, taking into account the axiom 2, it can be concluded that $f = \mathbf{D}^n F$ is a constant. Similarly it can be confirmed that the only distributions which have a derivative of order n equal to zero are the polynomials of degree less than n, and therefore that if $f, g \in \mathcal{C}_\infty$ the condition $\mathbf{D}^n f = \mathbf{D}^n g$ is equivalent to $f - g \in \mathcal{P}_n$.

We remark finally that from Theorem 1.1 it follows that if a distribution f is of degree $n > 0$ then each of its primitives is a distribution of degree $n - 1$.

1.6 Multiplicative Product.

As already remarked, the introduction of the concept of distribution has been motivated by the desire to extend the scope of the derivation operator, now understood in some generalised sense. However, as is not unusual in Mathematics, the gain from enlarging the set of mathematical objects in question is offset by the loss of some desirable properties.

Thus, so far as functions are concerned, the operation of multiplication is always possible and exhibits certain well-known characteristic properties. However, as we shall see, it is not possible to define the product of two arbitrary distributions so that all the desirable properties of multiplication continue to hold. Specifically we will show in the final part of this section that it is impossible to define a product of two arbitrary distributions which retains associativity, satisfies the usual derivation rule and coincides with the usual product in the case when both factors are continuous functions. There are several theories, of greater or lesser generality, for the multiplication of distributions and some of them even lead to the introduction of new mathematical objects.

Here we content ourselves with the simplest case in which the product fg is defined under the hypothesis that one of the factors is a function of class $\mathcal{C}^\infty$, while the other may be a distribution of arbitrary order. (Alternatively it may be assumed that one factor belongs to $\mathcal{C}^n$ while the other belongs to $\mathcal{C}_n$, for some $n \in \mathbb{N}$.) This new mode of multiplication may be introduced via the following Theorem:

Theorem 1.2 *There exists one and only one map $(\varphi, f) \rightsquigarrow \varphi f$, from $\mathcal{C}^\infty \times \mathcal{C}_\infty$ into $\mathcal{C}_\infty$, which satisfies the following conditions:*

p_1) *if $f \in \mathcal{C}$ then φf is the usual product of the functions φ, f;*

p_2) *for any $\varphi \in \mathcal{C}^\infty$ and any $f \in \mathcal{C}_\infty$, $\mathbf{D}(\varphi f) = \varphi' f + \varphi \mathbf{D} f$.*

Before we prove this theorem note that if there exists any map $(\varphi, f) \rightsquigarrow \varphi f$ for which condition p_2) holds then, by successive application of that property, in the form $\varphi \mathbf{D} f = \mathbf{D}(\varphi f) - \varphi' f$, we will also have that

$$
\begin{aligned}
\varphi \mathbf{D}^2 g &= \varphi \mathbf{D}\,(\mathbf{D} g) = \mathbf{D}\,(\varphi \mathbf{D} g) - \varphi' \mathbf{D} g \\
&= \mathbf{D}\,[\mathbf{D}\,(\varphi g) - \varphi' g] - [\mathbf{D}\,(\varphi' g) - \varphi'' g] \\
&= \mathbf{D}^2\,(\varphi g) - 2\mathbf{D}\,(\varphi' g) + \varphi'' g
\end{aligned}
$$

for any $\varphi \in \mathcal{C}^\infty$ and $g \in \mathcal{C}_\infty$.

We are therefore led to the following lemma, which will be proved before the theorem 1.2 itself:

Lemma 1.3 *If the mapping $(\varphi, f) \rightsquigarrow \varphi f$ satisfies the condition p_2) then for any $\varphi \in \mathcal{C}^\infty, g \in \mathcal{C}_\infty$ and $n \in \mathbb{N}$ we have*

$$
\varphi \mathbf{D}^n g = \sum_{k=0}^{n} (-1)^k \binom{n}{k} \mathbf{D}^{n-k}(\varphi^{(k)} g). \tag{1.1}
$$

Proof: The proof will proceed by induction on n. The result is trivially true for $n = 0$. Accordingly, suppose that it is true for some n and let $\varphi \in \mathcal{C}^\infty$, $g \in \mathcal{C}_\infty$. Taking the induction hypothesis, the property p_2) and the identity $\binom{n}{k} + \binom{n}{k-1} = \binom{n+1}{k}$ into account, we have:

$$
\begin{aligned}
\varphi \mathbf{D}^{n+1} g &= \varphi \mathbf{D}^n (\mathbf{D} g) \\
&= \sum_{k=0}^{n} (-1)^k \binom{n}{k} \mathbf{D}^{n-k}(\varphi^{(k)} \mathbf{D} g) \\
&= \sum_{k=0}^{n} (-1)^k \binom{n}{k} \mathbf{D}^{n-k+1}(\varphi^{(k)} g) - \sum_{k=0}^{n} (-1)^k \binom{n}{k} \mathbf{D}^{n-k}(\varphi^{(k+1)} g)
\end{aligned}
$$

and therefore,

$$
\begin{aligned}
\varphi \mathbf{D}^{n+1} g &= \mathbf{D}^{n+1}(\varphi g) + \sum_{k=1}^{n} (-1)^k \binom{n}{k} \mathbf{D}^{n-k+1}(\varphi^{(k)} g) \\
&\qquad - \sum_{k=1}^{n+1} (-1)^{k-1} \binom{n}{k-1} \mathbf{D}^{n-k+1}(\varphi^{(k)} g) \\
&= \sum_{k=0}^{n+1} (-1)^k \binom{n+1}{k} \mathbf{D}^{n+1-k}(\varphi^{(k)} g),
\end{aligned}
$$

which completes the proof. $\qquad\qquad\qquad\qquad\qquad\qquad\qquad\qquad\qquad\qquad\qquad\qquad\square$

Proof: (**Theorem 1.2**) We begin by showing the uniqueness of the mapping in question. To show that it is not possible to have more than one mapping satisfying the stated conditions it is enough to note that the formula

$$\varphi \mathbf{D} f \; = \; \mathbf{D}(\varphi f) - \varphi' f,$$

(which is a trivial consequence of $\mathbf{p}_2$)), implies that the product of the function φ with the derivative of the distribution f will be uniquely determined if the same is true of the products of φ and φ' with f. It follows that as soon as the sense attributed to the product of *any* function in $\mathcal{C}^\infty$ with a distribution f is fixed then the sense of the product of *any* function of class $\mathcal{C}^\infty$ by $\mathbf{D} f$ will clearly also be settled.

Hence, since by condition $\mathbf{p}_1$) the sense to be attributed to the product φf is uniquely determined in the case when $f \in \mathcal{C}$ and φ is an arbitrary element in $\mathcal{C}^\infty$, then the uniqueness of the product follows immediately by induction on the degree of the distribution f.

We now prove the existence of the mapping. According to the Lemma 1.3, if f is an arbitrary distribution and n an integer such that $f \in \mathcal{C}_n$ (that is, if there exists $F \in \mathcal{C}$ such that $f = \mathbf{D}^n F$) and if there does exist a mapping from $\mathcal{C}^\infty \times \mathcal{C}_\infty$ into $\mathcal{C}_\infty$ satisfying conditions $\mathbf{p}_1$) and $\mathbf{p}_2$), then we will necessarily have:

$$\varphi f \; = \; \varphi \mathbf{D}^n F \; = \; \sum_{k=0}^{n}(-1)^k \tbinom{n}{k}\mathbf{D}^{n-k}(\varphi^{(k)}F), \tag{1.2}$$

where the products $\varphi^{(k)} F$ may be understood in the usual sense.

Since all terms on right-hand side now have a precise sense it is only natural to use formula (1.2) to define the product φf and thereby to prove its existence.

First it is necessary to prove the coherence of the definition, that is, that if we have $\mathbf{D}^m F = \mathbf{D}^n G$ then we must have also

$$
\begin{aligned}
\varphi \mathbf{D}^m F \; &= \; \sum_{k=0}^{m}(-1)^k \tbinom{m}{k}\mathbf{D}^{m-k}(\varphi^{(k)}F) \\
&= \; \sum_{k=0}^{n}(-1)^k \tbinom{n}{k}\mathbf{D}^{n-k}(\varphi^{(k)}G) \; = \; \varphi \mathbf{D}^n G.
\end{aligned}
$$

For this purpose it is enough to show that the following two conditions are satisfied (all products of φ with the distributions therein being supposed defined by the formula (1.2)):

(1) - for any $\varphi \in \mathcal{C}^\infty$, $F \in \mathcal{C}$, $G \in \mathcal{C}$ and $n \in \mathbb{N}$ the equality

$$\mathbf{D}^n F \; = \; \mathbf{D}^n G$$

implies

$$\varphi \mathbf{D}^n F \; = \; \varphi \mathbf{D}^n G;$$

(2) - if $\varphi \in \mathcal{C}^\infty$, $F \in \mathcal{C}^1$ and $n \in \mathbb{N}$ then $\varphi \mathbf{D}^{n+1} F = \varphi \mathbf{D}^n F'$.

In fact suppose that both conditions are satisfied and let F, G, m and n be such that $F, G \in \mathcal{C}$ and $\mathbf{D}^m F = \mathbf{D}^n G$ where $m \le n$; also, let $\varphi \in \mathcal{C}^\infty$. Since $\mathbf{D}^m F = \mathbf{D}^n \mathcal{J}^{n-m} F$ then $\mathbf{D}^n \mathcal{J}^{n-m} F = \mathbf{D}^n G$ and therefore, according to condition **(1)**, $\varphi \mathbf{D}^n \mathcal{J}^{n-m} F = \varphi \mathbf{D}^n G$; hence by appealing to condition **(2)** we have immediately $\varphi \mathbf{D}^m F = \varphi \mathbf{D}^n G$, as asserted.

We now prove that conditions **(1)** and **(2)** do in fact hold. For condition **(1)** suppose that $\mathbf{D}^n F = \mathbf{D}^n G$, and therefore that $F = G + P$ with $P \in \mathcal{P}_n$. Then, taking into account the Leibniz formula and the fact that $P^{(n)} = 0$, we will get

$$
\begin{aligned}
\varphi \mathbf{D}^n F - \varphi \mathbf{D}^n G \;&=\; \sum_{k=0}^{n} (-1)^k \tbinom{n}{k} \mathbf{D}^{n-k}(\varphi^{(k)} F) \\[4pt]
&\quad - \sum_{k=0}^{n} (-1)^k \tbinom{n}{k} \mathbf{D}^{n-k}(\varphi^{(k)} G) \\[4pt]
&=\; \sum_{k=0}^{n} (-1)^k \tbinom{n}{k}(\varphi^{(k)} P)^{n-k} \;=\; \varphi P^{(n)} \;=\; 0.
\end{aligned}
$$

For condition **(2)** we have, under the hypothesis:

$$
\begin{aligned}
\varphi \mathbf{D}^n F' \;&=\; \sum_{k=0}^{n} (-1)^k \tbinom{n}{k} \mathbf{D}^{n-k}(\varphi^{(k)} F') \\[4pt]
&=\; \sum_{k=0}^{n} (-1)^k \tbinom{n}{k} \mathbf{D}^{n-k+1}(\varphi^{(k)} F) - \sum_{k=0}^{n} (-1)^k \tbinom{n}{k} \mathbf{D}^{n-k}(\varphi^{(k+1)} F) \\[4pt]
&=\; \mathbf{D}^{n+1}(\varphi F) + \sum_{k=1}^{n} (-1)^k \tbinom{n}{k} \mathbf{D}^{n-k+1}(\varphi^{(k)} F) \\[4pt]
&\quad - \sum_{k=1}^{n+1} (-1)^{k-1} \tbinom{n}{k-1} \mathbf{D}^{n-k+1}(\varphi^{(k)} F) \\[4pt]
&=\; \sum_{k=0}^{n+1} (-1)^k \tbinom{n+1}{k} \mathbf{D}^{n+1-k}(\varphi^{(k)} F) \;=\; \varphi \mathbf{D}^{n+1} F.
\end{aligned}
$$

We have therefore proved that formula (1.2) effectively defines a mapping from $\mathcal{C}^\infty \times \mathcal{C}_\infty$ into $\mathcal{C}_\infty$ and now it remains to prove that it satisfies properties $\mathbf{p_1})$ and $\mathbf{p_2})$. Property $\mathbf{p_1})$ is immediate. For property $\mathbf{p_2})$, let $\varphi \in \mathcal{C}^\infty$ and $f = \mathbf{D}^n F$, where $F \in \mathcal{C}$; then we have

$$
\begin{aligned}
\varphi' f + \varphi \mathbf{D} f \;&=\; \sum_{k=0}^{n} (-1)^k \tbinom{n}{k} \mathbf{D}^{n-k}(\varphi^{(k+1)} F) \\[4pt]
&\quad + \sum_{k=0}^{n+1} (-1)^k \tbinom{n+1}{k} \mathbf{D}^{n+1-k}(\varphi^{(k)} F) \\[4pt]
&=\; - \sum_{k=1}^{n+1} (-1)^k \tbinom{n}{k-1} \mathbf{D}^{n-k+1}(\varphi^{(k)} F) + \mathbf{D}^{n+1}(\varphi F) \\[4pt]
&\quad + \sum_{k=1}^{n+1} (-1)^k \tbinom{n+1}{k} \mathbf{D}^{n+1-k}(\varphi^{(k)} F) \\[4pt]
&=\; \mathbf{D}\left[\sum_{k=0}^{n} (-1)^k \tbinom{n}{k} \mathbf{D}^{n-k}(\varphi^{(k)} F) \right] \;=\; \mathbf{D}(\varphi \mathbf{D}^n F) \;=\; \mathbf{D}(\varphi f)
\end{aligned}
$$

which completes the proof. $\qquad\qquad\qquad\qquad\qquad\qquad\qquad\qquad\qquad\qquad\qquad\quad\square$

Before giving some examples it is convenient to state some very simple properties enjoyed by the product just defined, which is often referred to as the *Schwartz multiplicative product*.

Theorem 1.4 *Given $f, g \in \mathcal{C}_\infty$, $\varphi, \psi \in \mathcal{C}^\infty$ and $\alpha \in \mathbf{C}$, and denoting by* **0** *the zero function and by* **1** *the unity function (defined on the interval* **I** *under consideration) we have*

1. **0**$f = $ **0**, **1**$f = f$; *more generally, the product of the function that assumes the value α on each point of* **I** *by the distribution f is equal to the product of the complex number α by f;*

2. $(\varphi + \psi)f = \varphi f + \psi f$;

3. $\varphi(f + g) = \varphi f + \varphi g$;

4. $\varphi(\psi f) = (\varphi \psi)f$.

Proof: The first two properties are immediate consequences of formula (1.2); the third also follows from the same formula by representing f and g as derivatives of the same order of continuous functions. For the last property it is enough to observe that it is obviously true under the hypothesis that f is a continuous function (for any $\varphi, \psi \in \mathcal{C}^\infty$) and that, on the other hand, if it is satisfied for any distribution f (and for any $\varphi, \psi \in \mathcal{C}^\infty$) it will also be satisfied by $\mathbf{D}f$ as a consequence of the following relations:

$$\begin{aligned}
\varphi(\psi \mathbf{D}f) &= \varphi \mathbf{D}(\psi f) - \varphi(\psi' f) \\
&= \mathbf{D}[\varphi(\psi f)] - \varphi'(\psi f) - \varphi(\psi' f) \\
&= \mathbf{D}[(\varphi\psi)f] - (\varphi'\psi)f - (\varphi\psi')f \\
&= \mathbf{D}[(\varphi\psi)f] - (\varphi\psi)'f \ = \ (\varphi\psi)\mathbf{D}f.
\end{aligned}$$

$\square$

Throughout the following examples we allow the abuse of notation referred to in section 1.4, and we use the symbol $\delta(x)$ to denote the Dirac distribution, etc.

1. Suppose that the origin 0 belongs to the interior of the interval **I**; taking into account the equalities:

$$x\,\mathbf{H}(x) \ = \ x\,\mathbf{D}J(x) \ = \ \mathbf{D}[x\,J(x)] - J(x)$$

and the fact that $xJ(x)$ (product of a function of class $\mathcal{C}^\infty$ by a continuous function, therefore similar to the usual product) is a function of class $\mathcal{C}^1$, with derivative equal to $2J(x)$ (in the usual sense and therefore also in the sense of distributions) it can be concluded that

$$x\,\mathbf{H}(x) \ = \ J(x).$$

Differentiating both sides we get:

$$\mathbf{H}(x) + x\,\delta(x) \ = \ \mathbf{H}(x)$$

that is, $x\,\delta(x) = 0$.

For any integer $n > 1$ therefore we have

$$x^n \delta(x) \; = \; (x^{n-1} \cdot x)\delta(x) \; = \; x^{n-1}(x\,\delta(x)) \; = \; 0.$$

More generally, if m and n are non-negative integers, taking into account lemma 1.3, we obtain:

$$\begin{aligned}
x^m \delta^{(n)}(x) & = x^m \mathbf{D}^n \delta(x) \\
& = \sum_{k=0}^{n} (-1)^k \binom{n}{k} m(m-1)\cdots(m-k+1) \mathbf{D}^{n-k}\left(x^{m-k}\delta(x)\right).
\end{aligned}$$

If $m > n$, all products of the form $x^{m-k}\delta(x)$ are zero; if $m \le n$ the unique term in the summation which is not zero is that one which corresponds to $k = m$; hence,

$$x^m \delta^{(n)}(x) = \begin{cases} 0 & \text{if} \quad m > n \\ (-1)^m \dfrac{n!}{(n-m)!}\, \delta^{(n-m)}(x) & \text{if} \quad m \le n. \end{cases}$$

As the several cases considered in this example show, the degree of a product φf may be smaller that that of the distribution f; however, as an immediate consequence of formula (1.2), it is easy to see that it cannot be larger.

2. Suppose again that the origin belongs to the interior of the interval $\mathbf{I}$ and now let φ be a function in $\mathcal{C}^\infty(\mathbf{I})$ and n an arbitrary positive integer. Considering the MacLaurin formula for φ with integral remainder,

$$\varphi(x) \; = \; \varphi(0) + x\varphi'(0) + \cdots + \frac{x^n}{n!}\,\varphi^{(n)}(0) + \int_0^x \frac{(x-t)^n}{n!}\,\varphi^{(n+1)}(t)dt,$$

and making the change of variable $t = x\tau$, we get the formula

$$\varphi(x) \; = \; \sum_{k=0}^{n} \frac{x^k}{k!}\,\varphi^{(k)}(0) + \frac{x^{n+1}}{(n+1)!}\theta_{n+1}(x)$$

where

$$\theta_{n+1}(x) \; = \; (n+1)\int_0^1 (1-\tau)^n \varphi^{(n+1)}(x\tau)d\tau$$

is clearly a function in $\mathcal{C}^\infty$.

Taking into account the previous example we will get:

$$\begin{aligned}
\varphi(x)\delta^{(n)}(x) & = \sum_{k=0}^{n} \frac{\varphi^{(k)}(0)}{k!}\,x^k \delta^{(n)}(x) + \frac{1}{(n+1)!}\,\theta_{n+1}(x)\left[x^{n+1}\delta^{(n)}(x)\right] \\
& = \sum_{k=0}^{n} (-1)^k \binom{n}{k}\varphi^{(k)}(0)\delta^{(n-k)}(x).
\end{aligned}$$

In particular, for $n = 0$:

$$\varphi(x)\delta(x) \; = \; \varphi(0)\delta(x).$$

Later on we will see that the formulae stated here under the hypothesis that $\varphi \in \mathcal{C}^\infty$ are also valid under more general conditions.

3. We denote by λ the function defined and continuous on $\mathbb{R}$ such that

$$\lambda(x) \; = \; x \, \log |x| - x,$$

for each $x \neq 0$ and remark that the function $\omega : \mathbb{R} \to \mathbb{R}$, defined by $\omega(x) = x\lambda(x)$ belongs to $\mathcal{C}^1(\mathbb{R})$, which implies that its generalised derivative, $\mathbf{D}\omega$, coincides with its usual derivative ω'; hence

$$\mathbf{D}\omega(x) \; = \; 2\lambda(x) + x$$

and therefore

$$\mathbf{D}^2\omega(x) \; = \; 2\mathbf{D}\lambda(x) + 1. \tag{1.3}$$

However, from the product rule for differentiation we would get

$$\begin{aligned}
\mathbf{D}\omega(x) &= \mathbf{D}(x \, \lambda(x)) = \lambda(x) + x\mathbf{D}\lambda(x) \\
\mathbf{D}^2\omega(x) &= 2\mathbf{D}\lambda(x) + x\mathbf{D}^2\lambda(x),
\end{aligned}$$

and therefore, taking into account (1.3),

$$x\mathbf{D}^2\lambda(x) \; = \; 1.$$

Further still, taking into account the fact that $\mathbf{D}^2\lambda(x)$ is precisely that distribution which we have previously referred to as "the finite part of x^{-1}", we get

$$x \cdot \mathbf{Fp}\left\{\frac{1}{x}\right\} \; = \; 1.$$

Hence the distribution $\mathbf{Fp}\{x^{-1}\}$ behaves like a multiplicative inverse of the function x; moreover the same is true of any distribution of the form $\mathbf{Fp}\{x^{-1}\} + c\delta$, with $c \in \mathbf{C}$ since from the first of these examples it can be concluded that

$$x \left(\mathbf{Fp}\left\{\frac{1}{x}\right\} + c \, \delta(x)\right) \; = \; x\mathbf{Fp}\frac{1}{x} + cx\delta(x) \; = \; 1.$$

4. By induction on n (and using the rule for the differentiation of a product) it is easy to verify that, for any positive integers m and n we have:

$$x^m \mathbf{Fp}\left\{\frac{1}{x^n}\right\} = \begin{cases} x^{m-n} & \text{if } m \geq n \\ \mathbf{Fp}\{x^{-(n-m)}\} & \text{if } m < n. \end{cases}$$

We now register some remarks which are in some sense a complement to theorem 1.2. First note that a simple repetition of the proof of that theorem would allow us to prove the following proposition:

Proposition 1.5 *There exists one and only one map $(f, \varphi) \rightsquigarrow f\varphi$, from $\mathcal{C}_\infty \times \mathcal{C}^\infty$ into $\mathcal{C}_\infty$ such that $f\varphi$ coincides with the usual product if f is a continuous function and such that we have $\mathbf{D}(f\varphi) = (\mathbf{D}f)\varphi + f\varphi'$, for any $f \in \mathcal{C}_\infty$ and $\varphi \in \mathcal{C}^\infty$.*

Since the mapping that associates with the pair (f, φ) the product φf defined above satisfies those conditions the guarantee of uniqueness expressed in this proposition shows that we have necessarily $f\varphi = \varphi f$, for any $f \in \mathcal{C}_\infty$ and any $\varphi \in \mathcal{C}^\infty$.

Hence the product of two distributions f and g may always be formed whenever either of them is a function in $\mathcal{C}^\infty$, and it is commutative in the following sense: whenever either of the two products fg or gf exists so also does the other, and it has the same value. For example:

$$\delta(x)x \;=\; 0 \;, \quad \left(\mathbf{Fp}\left\{\frac{1}{x}\right\}\right) x \;=\; 1 \;, \quad \text{etc.}$$

Another fact which it is useful to mention is the following: if $f \in \mathcal{C}_n$, $f = \mathbf{D}^n F$ with $F \in \mathcal{C}$ then formula (1.2) allow us to determine the product φf (or $f\varphi$) without the need to assume that $\varphi \in \mathcal{C}^\infty$. In fact all terms in the right-hand side will have a precise sense provided that $\varphi \in \mathcal{C}^n$.

This observation leads immediately to an extension of the product previously considered, as stated in the following proposition (the proof of which is also trivially adapted from that of Theorem 1.2):

> **Proposition 1.6** *For each integer $n \geq 0$ it is possible (and in one way only) to associate with each pair $(\varphi, f) \in \mathcal{C}^n \times \mathcal{C}_n$ a distribution $\varphi f \in \mathcal{C}_n$, independent of n, and such that*
>
> $\mathbf{p}_1'$) *if $f \in \mathcal{C}$, φf is the usual product of φ and f;*
>
> $\mathbf{p}_2'$) $\mathbf{D}(\varphi f) = \varphi' f + \varphi \mathbf{D} f$, *for every $f \in \mathcal{C}_n$ and any $\varphi \in \mathcal{C}^{n+1}$.*

Even with this extension it is clear that the only functions which are multipliable by *any* distribution are the functions in $\mathcal{C}^\infty$.

This section will end with an example due to Laurent Schwartz which proves the impossibility of a general definition of a product of two arbitrary distributions. More rigorously, we will see that it does not exist a binary operation on $\mathcal{C}_\infty$, $(f, g) \rightsquigarrow f \bullet g$, which satisfies the following three conditions:

> **1.** if f and g are continuous functions $f \bullet g$ is the usual product;
>
> **2.** $\mathbf{D}(f \bullet g) = \mathbf{D} f \bullet g + f \bullet \mathbf{D} g$ for any $f, g \in \mathcal{C}_\infty$;
>
> **3.** $f \bullet (g \bullet h) = (f \bullet g) \bullet h$, for any $f, g, h \in \mathcal{C}_\infty$).

In fact, comparing the preceding conditions **1.** and **2.** with those demanded of the product in the statements of theorem 1.2 and of proposition 1.5, it is immediately clear that (if such an operation would exists) we should have for any $\varphi \in \mathcal{C}^\infty$ and $f \in \mathcal{C}_\infty$:

$$\varphi \bullet f \;=\; \varphi f \;=\; f \bullet \varphi.$$

In particular we would then have

$$\left(\mathbf{Fp}\left\{\frac{1}{x}\right\}\bullet x\right)\bullet\delta(x) \;=\; \left(\mathbf{Fp}\left\{\frac{1}{x}\right\}x\right)\bullet\delta(x)$$

$$=\; \mathbf{1}\bullet\delta(x) \;=\; 1\,\delta(x) \;=\; \delta(x)$$

and

$$\mathbf{Fp}\left\{\frac{1}{x}\right\}\bullet(x\bullet\delta(x)) \;=\; \mathbf{Fp}\left\{\frac{1}{x}\right\}\bullet(x\,\delta(x))$$

$$=\; \mathbf{Fp}\left\{\frac{1}{x}\right\}\bullet\mathbf{0} \;=\; \mathbf{Fp}\left\{\frac{1}{x}\right\}0 \;=\; \mathbf{0}$$

and therefore

$$\left(\mathbf{Fp}\left\{\frac{1}{x}\right\}\bullet x\right)\bullet\delta(x) \neq \mathbf{Fp}\left\{\frac{1}{x}\right\}\bullet(x\bullet\delta(x))$$

which contradicts condition **3**.

1.7   Change of Variable.

Except for certain particular cases, some of which nevertheless are of much interest in applications, the definition of the change of variable in the context of distribution theory is of considerable difficulty. Hence to begin with we content ourselves with the consideration of a very simple case.

If $\mathbf{I}$ and $\mathbf{J}$ are two intervals of $\mathrm{I\!R}$ and $\varphi : \mathbf{J} \to \mathbf{I}$ is a function which satisfies conditions which will be stated in the sequel, we will try to associate with each distribution f defined on $\mathbf{I}$ a distribution g defined on $\mathbf{J}$ which may reasonably be considered to be the composition of f with φ: $g = f \circ \varphi$. Essentially we shall demand that this composition should have the usual sense when f is a continuous function and that the usual chain rule for differentiation should be satisfied.

Let then $\mathbf{I}$ and $\mathbf{J}$ be two intervals of $\mathrm{I\!R}$ and suppose that $\varphi : \mathbf{J} \to \mathbf{I}$ is a continuous function. As is well-known a well defined mapping from $\mathcal{C}(\mathbf{I})$ into $\mathcal{C}(\mathbf{J})$ is then determined and this map associates with each function F, continuous on $\mathbf{I}$, its composition with φ, $F \circ \varphi$, which is a member of $\mathcal{C}(\mathbf{J})$. This mapping so defined is usually called "the composition of f with the function φ", or "the change of variable determined by the substitution $x = \varphi(t)$".

In particular, if $\varphi \in \mathcal{C}^1(\mathbf{J})$ and $F \in \mathcal{C}^1(\mathbf{I})$ then we also have $F \circ \varphi \in \mathcal{C}^1(\mathbf{J})$ and at each point $t \in \mathbf{J}$, the following equality

$$(F \circ \varphi)'(t) \;=\; F'(\varphi(t))\,\varphi'(t)$$

holds; that is

$$(F \circ \varphi)' \;=\; \varphi'(F' \circ \varphi).$$

Supposing additionally that $\varphi'(t) \neq 0$ for each $t \in \mathbf{J}$ (that is, that $\frac{1}{\varphi'} \in \mathcal{C}(\mathbf{J})$) we can also write

$$F' \circ \varphi \;=\; \frac{1}{\varphi'}\,(F \circ \varphi)'.$$

From now on it will be convenient for us to adopt a more explicit notation using the symbol $\mathbf{D}_t$ for the differentiation of functions in $\mathcal{C}^1(\mathbf{J})$ and the symbol $\mathbf{D}_x$ for the differentiation in $\mathcal{C}^1(\mathbf{I})$; the previous formula may therefore be written as

$$(\mathbf{D}_x F) \circ \varphi \;=\; \frac{1}{\varphi'} \, \mathbf{D}_t \left(F \circ \varphi \right),$$

or again, denoting by Δ_t the operator which to each function $G \in \mathcal{C}^1(\mathbf{J})$ makes correspond the function $\frac{1}{\varphi'} \, \mathbf{D}_t G$:

$$(\mathbf{D}_x F) \circ \varphi \;=\; \Delta_t \left(F \circ \varphi \right).$$

This relation may be expressed by saying that the change of variable determined by $x = \varphi(t)$ transforms the operator $\mathbf{D}_x$ into the operator Δ_t.

Supposing now that $\varphi \in \mathcal{C}^2(\mathbf{J})$ and $F \in \mathcal{C}^2(\mathbf{I})$, under the hypothesis that $\frac{1}{\varphi'} \in \mathcal{C}(\mathbf{J})$ we will have

$$\begin{aligned}
\left(\mathbf{D}_x^2 F\right) \circ \varphi \;&=\; \left[\mathbf{D}_x \left(\mathbf{D}_x F\right)\right] \circ \varphi \\
&=\; \Delta_t \left[(\mathbf{D}_x F) \circ \varphi\right] \;=\; \Delta_t \left[\Delta_t \left(F \circ \varphi\right)\right] \;=\; \Delta_t^2 \left(F \circ \varphi\right).
\end{aligned}$$

In general supposing that $1/\varphi' \in \mathcal{C}(\mathbf{J})$ and that $\varphi \in \mathcal{C}^n(\mathbf{J})$ and $F \in \mathcal{C}^n(\mathbf{I})$ it follows immediately by induction that

$$(\mathbf{D}_x^n F) \circ \varphi \;=\; \Delta_t^n \left(F \circ \varphi\right). \tag{1.4}$$

The results just obtained will now let us extend the change of variable to the space $\mathcal{C}_\infty(\mathbf{I})$. The extension will be based essentially on the following:

Theorem 1.7 *Let $\mathbf{I}$ and $\mathbf{J}$ be two intervals and let $\varphi : \mathbf{J} \to \mathbf{I}$ be a function in $\mathcal{C}^\infty$ which is such that $\varphi'(t) \neq 0$ for every $t \in \mathbf{J}$. Then there exists one and only one mapping $f \rightsquigarrow f \circ \varphi$ from $\mathcal{C}_\infty(\mathbf{I})$ into $\mathcal{C}_\infty(\mathbf{J})$ such that*

$\mathbf{c}_1$) *if $f \in \mathcal{C}(\mathbf{I})$ then $f \circ \varphi$ is the usual composition of f and φ;*

$\mathbf{c}_2$) *for any $f \in \mathcal{C}_\infty(\mathbf{I})$ the usual rule for differentiation holds, that is:[12]*

$$\mathbf{D}_t \left(f \circ \varphi\right) \;=\; \varphi' \left[(\mathbf{D}_x f) \circ \varphi\right].$$

Proof: The uniqueness is easily justified by induction on the degree of the distribution f, taking into account the fact that $\mathbf{c}_1$) determines the sense in which $f \circ \varphi$ must be understood when $f \in \mathcal{C}(\mathbf{I})$, and that the formula

$$(\mathbf{D}_x f) \circ \varphi \;=\; \frac{1}{\varphi'} \, \mathbf{D}_t \left(f \circ \varphi\right),$$

---

[12]The symbols $\mathbf{D}_x$ and $\mathbf{D}_t$ will from now on be understood to denote operators of generalised derivatives on $\mathcal{C}_\infty(\mathbf{I})$ and $\mathcal{C}_\infty(\mathbf{J})$, respectively.

an immediate consequence of $\mathbf{c_2}$), shows that the meaning of $(\mathbf{D}_x f) \circ \varphi$ will be fixed as soon as we fix the meaning of $f \circ \varphi$.

To prove existence we observe first that under the conditions of the hypothesis relating to φ, if f is an arbitrary distribution defined on $\mathbf{I}$, $f = \mathbf{D}^m F$ with $F \in \mathcal{C}(\mathbf{I})$, then the symbol[13]

$$\Delta_t^m(F \circ \varphi)$$

(where Δ_t now is to be understood as an operator defined in an obvious way on the space $\mathcal{C}_\infty(\mathbf{J})$) will always determine an element of $\mathcal{C}_\infty(\mathbf{I})$. However to see that a mapping from $\mathcal{C}_\infty(\mathbf{I})$ into $\mathcal{C}_\infty(\mathbf{J})$ is therefore defined it is still necessary to prove that this element in $\mathcal{C}_\infty(\mathbf{J})$, while naturally depending on f, is independent of the particular representation adopted for that distribution; that is, if $\mathbf{D}_x^m F = \mathbf{D}_x^n G$ then we also have

$$\Delta_t^m(F \circ \varphi) \;=\; \Delta_t^n(G \circ \varphi).$$

For this purpose it will be enough to prove that the following two conditions are satisfied:

1. for any $F, G \in \mathcal{C}(\mathbf{I})$ and any $n \in \mathbb{N}$ the equality $\mathbf{D}_x^n F = \mathbf{D}_x^n G$ implies $\Delta_t^n(F \circ \varphi) = \Delta_t^n(G \circ \varphi)$;

2. if $F \in \mathcal{C}^1(\mathbf{I})$ and $n \in \mathbb{N}$ then $\Delta_t^{n+1}(F \circ \varphi) = \Delta_t^n(\mathbf{D}_x F \circ \varphi)$.

To justify the first observe that if $\mathbf{D}_x^n F = \mathbf{D}_x^n G$, and therefore $F - G = P \in \mathcal{P}_n(\mathbf{I})$, then we also have

$$\Delta_t^n(F \circ \varphi) \;=\; \Delta_t^n(G \circ \varphi) + \Delta_t^n(P \circ \varphi) \;=\; \Delta_t^n(G \circ \varphi),$$

since by (1.4),

$$\Delta_t^n(P \circ \varphi) \;=\; (\mathbf{D}_x^n P) \circ \varphi \;=\; 0.$$

For the second condition it is enough to take into account that under the hypothesis we have again by (1.4):

$$\Delta_t^{n+1}(F \circ \varphi) \;=\; \Delta_t^n\left[\Delta_t(F \circ \varphi)\right] \;=\; \Delta_t^n(\mathbf{D}_x F \circ \varphi).$$

We have therefore proved that by making correspond to each distribution $f = \mathbf{D}_x^m F$, where $F \in \mathcal{C}(\mathbf{I})$, the distribution $\Delta_t^m(F \circ \varphi)$ we have really defined a mapping from $\mathcal{C}_\infty(\mathbf{I})$ into $\mathcal{C}_\infty(\mathbf{J})$. It remains to see if the map $f \rightsquigarrow f \circ \varphi$ defined in this way really satisfies the conditions $\mathbf{c_1}$) and $\mathbf{c_2}$). Since $\mathbf{c_1}$) is trivially satisfied, for $\mathbf{c_2}$) it is enough to observe that if we have $f = \mathbf{D}_x^m F$, we will also have $\mathbf{D}_x f = \mathbf{D}_x^{m+1} F$, and therefore

$$
\begin{aligned}
\varphi'\left[(\mathbf{D}_x f) \circ \varphi\right] \;&=\; \varphi' \Delta_t^{m+1}(F \circ \varphi) \\
&=\; \varphi' \frac{1}{\varphi'}\, \mathbf{D}_t\left[\Delta_t^m(F \circ \varphi)\right] \;=\; \mathbf{D}_t(f \circ \varphi).
\end{aligned}
$$

This completes the proof. $\square$

We may remark at this point that if $f = \mathbf{D}^n F \in \mathcal{C}_n(\mathbf{I})$ then for the second member of the equality

$$f \circ \varphi \;=\; \Delta_t^n(F \circ \varphi),$$

(which we have used to define $f \circ \varphi$) to determine effectively a distribution in $\mathcal{C}_\infty(\mathbf{J})$ it is not necessary to suppose that φ is a function in $\mathcal{C}^\infty$. It is enough to assume that

---

[13]The consideration of this symbol is naturally suggested by the formula (1.4).

$\varphi : \mathbf{J} \to \mathbf{I}$ satisfies the conditions that $\varphi \in \mathcal{C}^{n+1}(\mathbf{J})$ and $\varphi'(t) \neq 0$ for each $t \in \mathbf{J}$ (or, equivalently, that $\frac{1}{\varphi'} \in \mathcal{C}^n(\mathbf{J})$). Under these conditions a trivial adjustment of the preceding proof allows us to see that the following property holds.

Proposition 1.8 *Let* $\mathbf{I}$ *and* $\mathbf{J}$ *be two intervals; let* $f \in \mathcal{C}_n(\mathbf{I})$ *and* $\varphi : \mathbf{J} \to \mathbf{I}$ *be such that* $\varphi \in \mathcal{C}^n(\mathbf{J})$ *and (when* $n > 0$*)* $1/\varphi' \in \mathcal{C}^n(\mathbf{J})$. *Then it is possible to associate with the pair* (f, φ) *a uniquely defined distribution* $f \circ \varphi \in \mathcal{C}_n(\mathbf{J})$, *in such a way that we have*

$\mathbf{c_1'})$ *if* $f \in \mathcal{C}(\mathbf{I})$ *and* $\varphi \in \mathcal{C}(\mathbf{J})$, $f \circ \varphi$ *is the usual composition of the functions* f *and* φ;

$\mathbf{c_2'})$ *if* $f \in \mathcal{C}_n(\mathbf{I})$ *and* $\frac{1}{\varphi'} \in \mathcal{C}^{n+1}(\mathbf{J})$, $\mathbf{D}_t(f \circ \varphi) = \varphi'\,[(\mathbf{D}_x f) \circ \varphi]$.

The following theorem lists some important properties of the change of variable, in particular those related to operations on distributions defined earlier.

Theorem 1.9 *Let* $\mathbf{I}, \mathbf{J}$ *and* $\mathbf{K}$ *be three intervals of* $\mathbb{R}$, *and let* $\varphi : \mathbf{J} \to \mathbf{I}$ *and* $\psi : \mathbf{K} \to \mathbf{J}$ *be two functions of class* $\mathcal{C}^\infty$ *such that* $\varphi'(t) \neq 0$ *for each* $t \in \mathbf{J}$ *and* $\psi'(u) \neq 0$ *for each* $u \in \mathbf{K}$. *Also, let* $f, g \in \mathcal{C}_\infty(\mathbf{I})$, $\lambda \in \mathcal{C}^\infty(\mathbf{I})$ *and* $\alpha \in \mathbb{C}$. *Then we have,*

    i. $(f + g) \circ \varphi = f \circ \varphi + g \circ \varphi$;

    ii. $(\alpha f) \circ \varphi = \alpha(f \circ \varphi)$;

    iii. $(\lambda f) \circ \varphi = (\lambda \circ \varphi)(f \circ \varphi)$;

    iv. $(f \circ \varphi) \circ \psi = f \circ (\varphi \circ \psi)$.

Proof:

i. Representing f and g as derivatives of the same order of continuous functions on the interval $\mathbf{I}$, $f = \mathbf{D}_x^n F$, $g = \mathbf{D}_x^n G$ it follows at once that

$$
\begin{aligned}
(f + g) \circ \varphi &= [\mathbf{D}_x^n(F + G)] \circ \varphi = \Delta_t^n[(F + G) \circ \varphi] \\
&= \Delta_t^n(F \circ \varphi) + \Delta_t^n(G \circ \varphi) = f \circ \varphi + g \circ \varphi.
\end{aligned}
$$

ii. The proof of property **ii.** which, jointly with **i.**, expresses the linearity of the mapping $f \rightsquigarrow f \circ \varphi$, is also immediate.

iii. To prove **iii.** it is enough to verify the following facts:

    1) the property under consideration is true in the particular case that f is a continuous function (which is obvious);

    2) if the property is satisfied for a given distribution f (for any λ and φ under the stated conditions) the same will hold if we replace f by $\mathbf{D}f$.

Suppose then that the distribution f satisfies the condition

$$(\omega f) \circ \varphi = (\omega \circ \varphi)(f \circ \varphi),$$

for any $\omega \in \mathcal{C}^{\infty}(\mathbf{I})$. We would then have

$$
\begin{aligned}
(\lambda \mathbf{D}_x f) \circ \varphi &= [\mathbf{D}_x(\lambda f) - (\mathbf{D}_x \lambda)f] \circ \varphi \\
&= [\mathbf{D}_x(\lambda f)] \circ \varphi - [(\mathbf{D}_x \lambda)f] \circ \varphi \\
&= \Delta_t[(\lambda f) \circ \varphi] - [(\mathbf{D}_x \lambda)f] \circ \varphi,
\end{aligned}
$$

or, taking into account the condition supposed satisfied by f:

$$
\begin{aligned}
(\lambda \mathbf{D}_x f) \circ \varphi &= \Delta_t\left[(\lambda \circ \varphi)(f \circ \varphi)\right] - [(\mathbf{D}_x \lambda) \circ \varphi](f \circ \varphi) \\
&= \frac{1}{\varphi'}\left[(f \circ \varphi)\mathbf{D}_t(\lambda \circ \varphi) + (\lambda \circ \varphi)\mathbf{D}_t(f \circ \varphi)\right] \\
&\qquad - [(\mathbf{D}_x \lambda) \circ \varphi](f \circ \varphi) \\
&= (f \circ \varphi)[(\mathbf{D}_x \lambda) \circ \varphi] + (\lambda \circ \varphi)[(\mathbf{D}_x f) \circ \varphi] \\
&\qquad - (f \circ \varphi)[(\mathbf{D}_x \lambda) \circ \varphi] \\
&= (\lambda \circ \varphi)[(\mathbf{D}_x f) \circ \varphi].
\end{aligned}
$$

This completes the proof of **iii.**.

iv. Finally to prove **iv.** let $\theta = \varphi \circ \psi$; θ will be a map from $\mathbf{K}$ into $\mathbf{I}$, of class $\mathcal{C}^{\infty}$ and such that for any $u \in \mathbf{K}$, $\theta'(u) = (\varphi' \circ \psi)(u)\psi'(u) \neq 0$.

Under the conditions stated in Theorem 1.7 there will exist a *unique* map $f \rightsquigarrow f \circ \theta$, from $\mathcal{C}_{\infty}(\mathbf{I})$ into $\mathcal{C}_{\infty}(\mathbf{K})$, such that:

α) if $f \in \mathcal{C}(\mathbf{I})$, $f \circ \theta$ is the usual composed function of f and θ;

β) for any $f \in \mathcal{C}_{\infty}(\mathbf{I})$, $\mathbf{D}_u(f \circ \theta) = \theta'[(\mathbf{D}_x f) \circ \theta]$.

Hence, given the uniqueness of the mapping under consideration, to prove property **iv.** it will be enough to show that if we put $\Lambda(f) = (f \circ \varphi) \circ \psi$ for each $f \in \mathcal{C}_{\infty}(\mathbf{I})$ then the following conditions hold

α') for any $f \in \mathcal{C}(\mathbf{I})$, $\Lambda(f) = f \circ \theta$ (the usual composition of functions),

β') for any $f \in \mathcal{C}_{\infty}(\mathbf{I})$, $\mathbf{D}_u[\Lambda(f)] = \theta'\,\Lambda(\mathbf{D}_x f)$.

Condition α' is obvious. With respect to β', since the compositions with φ and ψ satisfy the derivation rule, we have (using property **iii.**),

$$
\begin{aligned}
\mathbf{D}_u[\Lambda(f)] &= \mathbf{D}_u[(f \circ \varphi) \circ \psi] = \psi'\{[\mathbf{D}_t(f \circ \varphi)] \circ \psi\} \\
&= \psi'\{[\varphi'(\mathbf{D}_x f \circ \varphi)] \circ \psi\} \\
&= \psi'(\varphi' \circ \psi)\{[(\mathbf{D}_x f) \circ \varphi] \circ \psi\} = \theta'\,\Lambda(\mathbf{D}_x f).
\end{aligned}
$$

The proof of the theorem is therefore complete. $\qquad\qquad\square$

We will end this section with some examples and complementary remarks in which it will be convenient to use symbols such as $f(\varphi(t))$ or $f(\varphi(x))$ to denote the composition $f \circ \varphi$ of the distribution f with the function φ. (This is strictly another abuse of notation, which nevertheless we will often resort to in the sequel).

1. Among the simplest examples of change of variable are the translations defined by substitutions of the form $x = t - a \; (a \in \mathbb{R})$.

If $\mathbf{I}$ is an interval and $F \in \mathcal{C}(\mathbf{I})$, it is usual to represent by $\tau_a F$ the function $F(\hat{t} - a)$ which is clearly continuous on the interval $\mathbf{J} = \{x + a : x \in \mathbf{I}\}$. The map τ_a that to each $F \in \mathcal{C}(\mathbf{I})$ makes correspond the function $\tau_a F \in \mathcal{C}(\mathbf{J})$ is called *the translation defined by the real number a*. The graph of $\tau_a F$ may be obtained from the graph of F by the geometric operation of translation of amplitude $|a|$, with the direction of the first coordinate axis in the positive or negative sense according as $a > 0$ or as $a < 0$.

In the case when $F \in \mathcal{C}^1(\mathbf{I})$ we will have $\tau_a F \in \mathcal{C}^1(\mathbf{J})$ and (using the same symbol to denote differentiation on $\mathcal{C}^1(\mathbf{I})$ or $\mathcal{C}^1(\mathbf{J})$):

$$\mathbf{D}\left(\tau_a F\right) \;=\; \tau_a(\mathbf{D}F).$$

Translation extends to distributions in the obvious way; if $f = \mathbf{D}^n F$, with $F \in \mathcal{C}(\mathbf{I})$ we have naturally:

$$\tau_a f \;=\; \tau_a\left(\mathbf{D}^n F\right) \;=\; \mathbf{D}^n\left(\tau_a F\right)$$

or, using a more suggestive notation:

$$f(t - a) \;=\; \tau_a f(t) \;=\; \mathbf{D}^n F(t - a).$$

For example, if $\mathbf{I} = \mathbb{R}$ and a is an arbitrary real number it is usual to denote by $\delta_{(a)}$ the distribution obtained from δ by the translation τ_a:

$$\delta_{(a)}(t) \;=\; \delta(t - a) \;=\; \mathbf{D}^2 J(t - a).$$

More generally, for each $p \in \mathbb{N}$:

$$\delta_{(a)}^{(p)}(t) \;=\; \delta^{(p)}(t - a) \;=\; \mathbf{D}^{p+2} J(t - a).$$

In section 1.6 we verified that, for every $n \in \mathbb{N}$ and any $\theta \in \mathcal{C}^\infty(\mathbb{R})$, the following formula

$$\theta(x)\delta^{(n)}(x) \;=\; \sum_{k=0}^{n}(-1)^k \tbinom{n}{k}\theta^{(k)}(0)\delta^{(n-k)}(x)$$

holds. If we make the change of variable $x = t - a$ (and take into account some of the properties expressed in theorem 1.9) we get:

$$\theta(t - a)\delta^{(n)}(t - a) \;=\; \sum_{k=0}^{n}(-1)^k \tbinom{n}{k}\theta^{(k)}(0)\delta^{(n-k)}(t - a).$$

Hence, if $\omega \in \mathcal{C}^\infty(\mathbb{R})$ is such that $\theta(t) = \omega(t + a)$ then we also have:

$$\omega(t)\delta^{(n)}(t - a) \;=\; \sum_{k=0}^{n}(-1)^k \tbinom{n}{k}\omega^{(k)}(a)\delta^{(n-k)}(t - a).$$

In particular, for $n = 0$:

$$\omega(t)\delta(t - a) \;=\; \omega(a)\delta(t - a).$$

2. More generally, let us consider the interpretation of $\delta(\varphi(t))$ when $\varphi : \mathbb{R} \to \mathbb{R}$ is any function in $\mathcal{C}^\infty(\mathbb{R})$ such that $\varphi'(t) \neq 0$ for each $t \in \mathbb{R}$. It is clear that such a function φ cannot have more than one real root and if it has none then all values of $\varphi(t)$ have the same sign. Let us make the change of variable $x = \varphi(t)$ in both members of the equality

$$J(x) \;=\; x\,\mathbf{H}(x),$$

obtained in section 1.6. We then get

$$J(\varphi(t)) \;=\; \varphi(t)\,\mathbf{H}(\varphi(t)).$$

If we have $\varphi(t) < 0$ [respectively $\varphi(t) > 0$] for every $t \in \mathbb{R}$ then we will have $J(\varphi(t)) = 0$ [respectively $J(\varphi(t)) = \varphi(t)$] and therefore, by derivation, $\mathbf{H}(\varphi(t)) = 0$, $\delta(\varphi(t)) = 0$ [respectively $\mathbf{H}(\varphi(t)) = 1$, $\delta(\varphi(t)) = 0$].

The more interesting case arise when there exists $a \in \mathbb{R}$ such that $\varphi(a) = 0$. We will then have, under the hypothesis that $\varphi'(t) > 0$,

$$J(\varphi(t)) \;=\; \varphi(t)\,\mathbf{H}(t - a)$$

and therefore

$$\varphi'(t)\,\mathbf{H}(\varphi(t)) \;=\; \varphi'(t)\,\mathbf{H}(t - a) + \varphi(t)\delta(t - a)$$

or, since $\varphi(a) = 0$:

$$\mathbf{H}(\varphi(t)) \;=\; \mathbf{H}(t - a) + \frac{\varphi(a)}{\varphi'(a)}\,\delta(t - a) \;=\; \mathbf{H}(t - a).$$

Hence we get

$$\delta(\varphi(t)) \;=\; \frac{1}{\varphi'(t)}\,\delta(t - a) \;=\; \frac{1}{\varphi'(a)}\,\delta(t - a).$$

On the other hand, under the hypothesis that $\varphi'(t) < 0$ we get

$$J(\varphi(t)) \;=\; \varphi(t)[1 - \mathbf{H}(t - a)],$$

and therefore

$$\mathbf{H}(\varphi(t)) \;=\; 1 - \mathbf{H}(t - a)$$

so that

$$\delta(\varphi(t)) \;=\; -\frac{1}{\varphi'(a)}\,\delta(t - a).$$

Both cases are included in the formula,

$$\delta(\varphi(t)) \;=\; \frac{1}{|\varphi'(a)|}\,\delta(t - a).$$

Hence, for example, if $\alpha \neq 0$ we will have:

$$\delta(\alpha t) \;=\; \frac{1}{|\alpha|}\,\delta(t),$$

from which by derivation we obtain:

$$\delta^{(p)}(\alpha t) \;=\; \frac{1}{|\alpha|\alpha^p}\,\delta^{(p)}(t),$$

for each $p \in \mathbb{N}$. In particular, for $\alpha = -1$:

$$\delta^{(p)}(-t) \;=\; (-1)^p \delta^{(p)}(t).$$

3. Consider next the corresponding problem posed by the expression $\mathbf{Fp}\{1/\varphi(t)\}$. Again let φ be a function in $\mathcal{C}^\infty(\mathbb{R})$ such that $\varphi'(t) \neq 0$ for each $t \in \mathbb{R}$.

Substituting x for $\varphi(t)$ in both members of the equality

$$\mathbf{Fp}\left\{\frac{1}{x}\right\} \;=\; \mathbf{D}_x \log |x|,$$

we will obtain

$$\mathbf{Fp}\left\{\frac{1}{\varphi(t)}\right\} \;=\; \Delta_t \log|\varphi(t)| \;=\; \frac{1}{\varphi'(t)}\,\mathbf{D}_t \log|\varphi(t)|.$$

If $\varphi(t)$ is never zero we obtain immediately

$$\mathbf{Fp}\left\{\frac{1}{\varphi(t)}\right\} \;=\; \frac{1}{\varphi'(t)}\,\frac{\varphi'(t)}{\varphi(t)} \;=\; \frac{1}{\varphi(t)}$$

On the other hand, if $\varphi(t)$ has a real root a, we have

$$\varphi(t) \;=\; (t-a)\,\theta(t),$$

where

$$\theta(t) \;=\; \int_0^1 \varphi'(a + \xi(t-a))d\xi$$

is a function of the class $\mathcal{C}^\infty$ and of constant sign (equal to the sign of φ'). From this it follows that

$$\mathbf{Fp}\left\{\frac{1}{\varphi(t)}\right\} \;=\; \frac{1}{\varphi'(t)}\,\mathbf{D}_t\,[\,\log|t-a| + \log|\theta(t)|\,]$$

or

$$\mathbf{Fp}\left\{\frac{1}{\varphi(t)}\right\} \;=\; \frac{1}{\varphi'(t)}\left(\mathbf{Pf}\left\{\frac{1}{t-a}\right\} + \frac{\theta'(t)}{\theta(t)}\right)$$

$$\;=\; \frac{1}{\varphi'(t)}\left(\mathbf{Pf}\left\{\frac{1}{t-a}\right\} + \mathbf{D}_t \log\left|\frac{\varphi(t)}{t-a}\right|\right).$$

If $\alpha \neq 0$ we have, for example,

$$\mathbf{Fp}\left\{\frac{1}{\alpha t}\right\} \;=\; \frac{1}{\alpha}\,\mathbf{Fp}\left\{\frac{1}{t}\right\}$$

and, in particular, for $\alpha = -1$:

$$\mathbf{Fp}\left\{-\frac{1}{t}\right\} = -\mathbf{Fp}\left\{\frac{1}{t}\right\}.$$

4. In what follows we always suppose that $f \in \mathcal{C}_\infty(\mathbb{R})$.

If $a \in \mathbb{R}$ we will say that *f has a period a* or that *a is a period of f* if and only if f is invariant for the translation τ_a, that is, if $\tau_a f = f$ (in other words, if $f(x-a) = f(x)$). A distribution f is said to be *periodic* if and only if it has a period $a \neq 0$. The continuous periodic functions (in the usual sense) are trivial examples of periodic distributions.

Let $\sigma(x)$ denote the characteristic part of the real number x; that is

$$\sigma(x) = \max\{z \in \mathbb{Z} : z \leq x\}.$$

The function σ, whose graph takes the form of a regular, ascending *"staircase"*, will have an indefinite integral over $\mathbb{R}$ and may therefore be identified with a distribution. Further, $\mathbf{D}\sigma$ will be a periodic distribution, with period 1; this can be seen immediately on differentiating both members of the identity:

$$\sigma(x) - \sigma(x-1) = 1.$$

Next, let $\gamma \in \mathbb{C}$. The distribution $f \in \mathcal{C}_\infty(\mathbb{R})$ is said to be *homogeneous of degree* γ if, for any $\alpha > 0$, the equality $f(\alpha x) = \alpha^\gamma f(x)$ holds. In the above examples we have already seen that the distribution $\delta^{(p)}$ is homogeneous of degree $-(p+1)$ and that $\mathbf{Fp}\{x^{-1}\}$ is homogeneous of degree -1.

Finally we say that f is an *even* [respectively an *odd*] distribution if $f(-x) = f(x)$ [respectively $f(-x) = -f(x)$]; $\delta^{(p)}$ is even or odd according as to whether p is even or odd; $\mathbf{Fp}\{x^{-1}\}$ is odd and is precisely the only odd distribution for which the product by x is equal to unity.

1.8   Restriction and Extension. The Division Problem.

Let $\mathbf{I}$ and $\mathbf{J}$ be two intervals of $\mathbb{R}$ such that $\mathbf{J} \subset \mathbf{I}$. If $F \in \mathcal{C}(\mathbf{I})$ the *restriction* of F to $\mathbf{J}$ is the function $F_{|\mathbf{J}}$ defined by

$$F_{|\mathbf{J}}(t) = F(t)$$

for each $t \in \mathbf{J}$. Clearly $F_{|\mathbf{J}} \in \mathcal{C}(\mathbf{J})$.

It is obvious that if $F \in \mathcal{C}^1(\mathbf{I})$ then we also have $F_{|\mathbf{J}} \in \mathcal{C}^1(\mathbf{J})$ and, moreover

$$(\mathbf{D}F)_{|\mathbf{J}} = \mathbf{D}\left(F_{|\mathbf{J}}\right),$$

that is, differentiation and restriction are commutative operations.

It is therefore natural, given a distribution $f = \mathbf{D}^n F$, with $F \in \mathcal{C}(\mathbf{I})$, to call the distribution denoted by $f_{|\mathbf{J}}$ and defined by $f_{|\mathbf{J}} = \mathbf{D}^n F_{|\mathbf{J}}$ *the restriction of f to the interval* $\mathbf{J}$.

To denote the restriction of f to $\mathbf{J}$ it will be useful to introduce the alternative notation $\rho_{\mathbf{J}}(f)$, in which the symbol $\rho_{\mathbf{J}}$ denotes the restriction operator, that is, the mapping $f \rightsquigarrow f_{|\mathbf{J}}$ from $\mathcal{C}_\infty(\mathbf{I})$ into $\mathcal{C}_\infty(\mathbf{J})$ that to each $f \in \mathcal{C}_\infty(\mathbf{I})$ makes correspond its restriction to $\mathbf{J}$. It is then obvious that for any $f \in \mathcal{C}_\infty(\mathbf{I})$ we have:

$$\rho_{\mathbf{J}}(\mathbf{D}f) = \mathbf{D}(\rho_{\mathbf{J}}f).$$

For example, with $\mathbf{I} = \mathbb{R}$ and $\mathbf{J} = [0, +\infty[$ it can be seen immediately that we have $\delta_{|\mathbf{J}} = \mathbf{0}$ and that the restriction of any distribution on $\mathbb{R}$ of the form $\delta_{(a)}^{(p)}$ to any one of the intervals $]-\infty, a]$ and $[a, +\infty[$ (or any subinterval of them) is also the null distribution. However, the restriction of $\delta_{(a)}^{(p)}$ to any interval which contains the point a in its interior is not the null distribution on that interval. In this case it is usual to continue to use the same symbol $\delta_{(a)}^{(p)}$ for this restriction, whenever no confusion is likely to arise, and we shall generally do so here.

As some of these examples show the operation of restriction may lower the degree of a distribution; it is obvious, however, that the degree will never increase in any case.

The restriction is a particular case of the operation of composition: if, given two intervals $\mathbf{I}$ and $\mathbf{J}$ such that $\mathbf{J} \subset \mathbf{I}$, we denote by $\imath$ the *canonical injection* of $\mathbf{J}$ into $\mathbf{I}$ (that is, the mapping $\imath : \mathbf{J} \to \mathbf{I}$ defined by $\imath(t) = t$ for each $t \in \mathbf{J}$) we will have for any distribution $f \in \mathcal{C}_\infty(\mathbf{I})$, $f_{|\mathbf{J}} = f \circ \imath$.

From this fact and from Theorem 1.9 the properties stated in the following theorem are an immediate consequence

> **Theorem 1.10** *Let* $\mathbf{I}$, $\mathbf{J}$, $\mathbf{K}$ *be three intervals such that* $\mathbf{K} \subset \mathbf{J} \subset \mathbf{I}$, f *and* g *two distributions in* $\mathcal{C}_\infty(\mathbf{I})$, λ *a function in* $\mathcal{C}^\infty(\mathbf{I})$ *and* $\alpha \in \mathbf{C}$. *Then we have*
>
> i) $(f + g)_{|\mathbf{J}} = f_{|\mathbf{J}} + g_{|\mathbf{J}}$;
>
> ii) $(\alpha f)_{|\mathbf{J}} = \alpha f_{|\mathbf{J}}$;
>
> iii) $(\lambda f)_{|\mathbf{J}} = \lambda_{|\mathbf{J}} f_{|\mathbf{J}}$;
>
> iv) $(f_{|\mathbf{J}})_{|\mathbf{K}} = f_{|\mathbf{K}}$.

The concept of restriction allow us to define in more general situations some of the algebraic operations which are basic in the distribution setting. We will mention, as an example, the case of addition for which we could now adopt the following definition: If $\mathbf{I}$ and $\mathbf{J}$ are two intervals of $\mathbb{R}$ with non-degenerate intersection $\mathbf{K} = \mathbf{I} \cap \mathbf{J}$ and if $f \in \mathcal{C}_\infty(\mathbf{I})$ and $g \in \mathcal{C}_\infty(\mathbf{J})$, we call the *sum of f and g* the distribution defined by:

$$f + g = \rho_{\mathbf{K}}(f) + \rho_{\mathbf{K}}(g)$$

(where the addition referred to in the right-hand side is the operation defined above in $\mathcal{C}_\infty(\mathbf{K})$).

Another notion which we can introduce naturally by making appeal to the concept of restriction is that of equality of two distributions on an interval which is contained within the domains of both distributions: if $\mathbf{I}, \mathbf{J}, \mathbf{K}$ are three (non-degenerate) intervals such that $\mathbf{K} \subset \mathbf{J} \cap \mathbf{I}$, we will say that the distributions $f \in \mathcal{C}_\infty(\mathbf{I})$ and $g \in \mathcal{C}_\infty(\mathbf{J})$ *are equal on* $\mathbf{K}$, and we write $f = g$ on $\mathbf{K}$, if the restrictions of f and g to the interval $\mathbf{K}$ are equal. Hence, for example, if $\mathbf{K}$ is a non-degenerate interval for which the point a is not interior we have that $\delta_{(a)}^{(p)} = \mathbf{0}$ in $\mathbf{K}$ for every $p \in \mathbb{N}$.

We now prove a result which will be useful in the sequel:

> **Theorem 1.11** *Let* $\mathbf{I}$ *be a non-degenerate interval of* $\mathbb{R}$ *and* f *and* g *distributions defined on* $\mathbf{I}$; *if, for every compact interval* $\mathbf{K} \subset \mathbf{I}$, *we have* $f = g$ *on* $\mathbf{K}$ *then the equality* $f = g$ *holds.*

Proof: The theorem will be proved provided we show that whenever we have $f = \mathbf{0}$ on $\mathbf{K}$, for every compact interval $\mathbf{K} \subset \mathbf{I}$, then $f = \mathbf{0}$. The theorem then follows immediately on considering the difference $f - g$. In the proof that follows we will suppose that the interval $\mathbf{I}$ is open (bounded or not); the study of the other cases is left as an exercise.

Let a_n and b_n be two sequences of points in $\mathbf{I}$, the first converging decreasingly to the lower bound of $\mathbf{I}$ and the second converging increasingly to the upper bound of $\mathbf{I}$; moreover suppose that $a_1 < b_1$ and, for each positive integer n, set $\mathbf{K}_n = [a_n, b_n]$. Then we will have

$$\mathbf{K}_1 \subset \mathbf{K}_2 \subset \cdots \qquad \text{and} \qquad \bigcup_{n \geq 1} \mathbf{K}_n = \mathbf{I}.$$

Further, let $f = \mathbf{D}^p F$ with $p \in \mathbb{N}$ and $F \in \mathcal{C}(\mathbf{I})$.

On the interval $\mathbf{K}_1$ the function F will be equal to a polynomial P_1 of degree $< p$ (since $\mathbf{D}^p F = 0$ on $\mathbf{K}_1$). We will show now that we must have $F(\alpha) = P_1(\alpha)$ for every $\alpha \in \mathbf{I}$, and this will complete the proof.

If α is any point in $\mathbf{I}$ let $j \in \mathbb{N}$ be such that $\alpha \in \mathbf{K}_j$; by the same reasoning there must exist a polynomial P_j of degree $< p$ such that $F = P_j$ on $\mathbf{K}_j$ and therefore $F(\alpha) = P_j(\alpha)$. However, since P_1 is the restriction of P_j to the interval $\mathbf{K}_1$, it may be concluded that the two polynomials P_1 and P_j are identical and therefore it follows that $F(\alpha) = P_1(\alpha)$. $\square$

Consider again two intervals $\mathbf{I}$ and $\mathbf{J}$ such that $\mathbf{J} \subset \mathbf{I}$ and let $f \in \mathcal{C}_\infty(\mathbf{I})$ and $g \in \mathcal{C}_\infty(\mathbf{J})$. It will be said that f is an *extension* of g (or, more precisely, an *extension of* g *to* $\mathbf{I}$) if $f_{|\mathbf{J}} = g$. When there exists an extension of the distribution g to the interval $\mathbf{I}$ then g is said to be *prolongable to* $\mathbf{I}$.

While the restriction operation (to a non-degenerate interval contained in the domain of the distribution considered) is an operation which is always possible and which has a unique result, the same cannot be said with respect to the extension operation. This is not always possible and, even when it can be carried out, the result is generally indeterminate.

Consider first the case of a distribution f defined on an interval $\mathbf{I} = [a, b]$, with $a, b \in \mathbb{R}$ and $a < b$. Representing f as $f = \mathbf{D}^n F$ with $F \in \mathcal{C}(\mathbf{I})$ and denoting by $\tilde{F}$ an arbitrarily chosen continuous function on $\mathbb{R}$ such that $\tilde{F}_{|\mathbf{I}} = F$, it is obvious that the distribution $\tilde{f} = \mathbf{D}^n \tilde{F}$ will be an extension of f to $\mathbb{R}$ and that if we want an extension to another interval $\mathbf{I}^* \supset \mathbf{I}$ it will be enough to restrict $\tilde{f}$ to $\mathbf{I}^*$. It is also clear that in this case there always exist infinitely many different extensions (except for the case $\mathbf{I}^* = \mathbf{I}$).

In the same way it is possible to see that any distribution defined on an interval of the form $]-\infty, a]$ (or $[a, +\infty[$) with $a \in \mathbb{R}$ is always prolongable in infinitely many different ways to any interval which contains strictly the domain of the distribution under consideration.

The situation is clearly different when the interval in which the distribution is defined has an end-point which does not belong to it. Taking as a paradigm the case $\mathbf{I} =]-\infty, a[$ it is easy to see that the following proposition holds:

> **Proposition 1.12** *Let $a \in \mathbb{R}$, $\mathbf{I} =]-\infty, a[$ and $f \in \mathcal{C}_\infty(\mathbf{I})$. For an extension of f to $\bar{\mathbf{I}} =]-\infty, a]$ (and therefore to $\mathbb{R}$) to exist it is necessary and sufficient that there exists a natural number n and a function F, which is continuous on $\mathbf{I}$ and bounded on the intersection of $\mathbf{I}$ with a neighbourhood of a, such that $f = \mathbf{D}^n F$.*

Proof: The condition is clearly necessary; supposing that it holds, to obtain an extension g of f to $\mathbb{R}$ it will be enough to choose arbitrarily a function $G \in \mathcal{C}(\mathbb{R})$ such that for $x < a$

$$G(x) \;=\; \int_a^x F(t)dt$$

and then set $g = \mathbf{D}^{n+1}G$.   □

We state the following proposition for future reference:

> **Proposition 1.13** *If $f \in \mathcal{C}(]-\infty, a[)$ extends to $\mathbb{R}$ then the same holds for its derivative and for any distribution of the form $(a - x)^{-p} f(x)$ with $p \in \mathbb{N}$.*

Proof: The proposition 1.12 makes it clear that, if f is prolongable, then the same holds for the derivatives of f of all orders.

To prove the final assertion in the statement it is enough to show that if f extends then so also does $(a - x)^{-1} f(x)$. If now $f = \mathbf{D}^n F$ with F continuous on $\mathbf{I} =]-\infty, a[$ and bounded on the intersection of $\mathbf{I}$ with a neighbourhood of a, then the distribution $(a - x)^{-1} f(x)$ may be expressed as a linear combination of the derivatives of functions of the form

$$G(x) \;=\; \frac{1}{(a - x)^k}\, F(x) \tag{1.5}$$

with $k = 1, 2, \ldots, n + 1$.

Hence the proof will be completed when we have shown that any function of the form (1.5) may be represented as a derivative, of an appropriate order, of a function continuous on $\mathbf{I}$ and bounded on the intersection of this interval with a neighbourhood of the point a.

To that end, supposing that $|F(x)| \leq M$ for $x \in [c, a[$ with $c < a$, consider the function defined on $\mathbf{I}$ by the formula:

$$\Phi(x) = \int_c^x \frac{(x-t)^k}{k!}\, G(t)dt = \frac{1}{k!}\int_c^x \left(\frac{x-t}{a-t}\right)^k F(t)dt.$$

This is a primitive of order $k+1$ of the function G, more precisely that primitive which is zero, together with all its derivatives of order $\leq k$, at the point c. In fact the preceding formula is nothing but the Taylor formula with remainder in the integral form for the function Φ. For any $x \in [c, a[$ we have:

$$\begin{aligned}
|\Phi(x)| &\leq \frac{1}{k!}\int_c^x \left(\frac{x-t}{a-t}\right)^k |F(t)|dt \\
&\leq \frac{1}{k!}\int_c^x |F(t)|dt \leq \frac{M}{k!}\,(x-c) \leq \frac{M}{k!}\,(a-c)
\end{aligned}$$

and the proof is complete. $\qquad\square$

It is also easily deduced from proposition 1.12 that in the case of an interval $\mathbf{I}$ of the form considered (or, of any other interval without an endpoint) there are distributions - and continuous functions - defined on $\mathbf{I}$ that are not prolongable to $\mathbb{R}$ (as distributions). This is what occurs with any function[14] $G \in \mathcal{C}(\mathbf{I})$ such that

$$\lim_{x \to a}(a - x)^p G(x) = +\infty, \tag{1.6}$$

for every $p \in \mathbb{N}$. In fact, by appealing to one of the usual rules for the calculation of limits, it is easily concluded that if the function G satisfies (1.6) each one of its primitives (of any order) satisfies the same condition. Hence G is not prolongable to $\mathbb{R}$, as a distribution.

Consider now another type of problem of extension of distributions which will be of interest in the sequel. We may state it in the following terms:

If $a \in \mathbb{R}$ set $\mathbf{I}_1 =] - \infty, a[$, $\mathbf{I}_2 =]a, +\infty[$ and let $f_1 \in \mathcal{C}_\infty(\mathbf{I}_1)$, $f_2 \in \mathcal{C}_\infty(\mathbf{I}_2)$. Under what conditions can we guarantee the existence of *a common extension* for f_1 and f_2 to $\mathbb{R}$, that is, of a distribution $f \in \mathcal{C}_\infty(\mathbb{R})$ such that $f_{|\mathbf{I}_1} = f_1$ and $f_{|\mathbf{I}_2} = f_2$? Moreover, when such an extension does exist will it be unique?

It is clear that an extension can exist only if both f_1 and f_2 may be extended to $\mathbb{R}$ (or, which is the same thing, if f_1 may be extended to $\bar{\mathbf{I}}_1$ and f_2 to $\bar{\mathbf{I}}_2$). By making appeal to proposition 1.12 it is easy to see that for this to hold there must exist a natural number n and two functions $F_1 \in \mathcal{C}(\mathbf{I}_1)$ and $F_2 \in \mathcal{C}(\mathbf{I}_2)$, both bounded on the intersection of their respective domains with a neighbourhood of the point a, such that $f_1 = \mathbf{D}^n F_1$ and $f_2 = \mathbf{D}^n F_2$. However, if this condition is satisfied, it will be enough to put

$$G_1(x) = \int_a^x F_1(t)dt \quad \text{if } x < a$$

---

[14]As an example, with $\mathbf{I} =] - \infty, a[$, take $G(x) = \exp\left(\frac{1}{a-x}\right)$.

$$G_2(x) \; = \; \int_a^x F_2(t)\,dt \quad \text{if} \quad x > a$$

and finally

$$G(x) = \begin{cases} G_1(x) & \text{if} \quad x < a \\ 0 & \text{if} \quad x = a \\ G_2(x) & \text{if} \quad x > a \end{cases}$$

for the distribution $f = \mathbf{D}^{n+1}G$ to be a common extension to $\mathbb{R}$ of f_1 and f_2.

We can therefore conclude that the condition for f_1 and f_2 to be prolongable to $\mathbb{R}$ is not only necessary but also sufficient to guarantee the existence of a common extension.

And what about uniqueness? Will the extension obtained under the hypothesis referred to be unique?

We shall see that the answer to this question is always negative.

If f and g are two common extensions of f_1 and f_2 to $\mathbb{R}$ then $g - f$ will be a distribution defined on $\mathbb{R}$ and with null restriction to each of the intervals $\mathbf{I}_1$ and $\mathbf{I}_2$. In the inverse sense, if once having obtained an extension f we add to it a distribution defined on $\mathbb{R}$ and null on $\mathbf{I}_1$ and $\mathbf{I}_2$ we will obtain a new extension of f_1 and f_2 to $\mathbb{R}$.

It is therefore of interest to determine which are the distributions defined on $\mathbb{R}$ which are null both in $\mathbf{I}_1$ and in $\mathbf{I}_2$ since the common extension of f_1 and f_2 is unique except for a distribution of that type.

Simple examples of distributions which are null on $\mathbf{I}_1$ and on $\mathbf{I}_2$ are the Dirac distribution at the point a and its derivatives; more generally, any linear combination of derivatives of $\delta_{(a)}$, that is, any distribution of the form

$$\sum_{i=0}^{n} \alpha_i \delta_{(a)}^{(i)},$$

with $n \in \mathbb{N}$ and $\alpha_i \in \mathbf{C}$, is zero on $\mathbf{I}_1$ and $\mathbf{I}_2$.

The following theorem shows that reciprocally any distribution that is zero on $\mathbf{I}_1$ and $\mathbf{I}_2$ is a linear combination of derivatives of $\delta_{(a)}$.

Theorem 1.14 *Let $f \in \mathcal{C}_n(\mathbb{R})$ with $n \geq 2$ and suppose that $f_{|\mathbf{I}_1} = \mathbf{0}$ and $f_{|\mathbf{I}_2} = \mathbf{0}$ (where $\mathbf{I}_1 =] - \infty, a[$ and $\mathbf{I}_2 =]a, +\infty[$); then there exist constants $\alpha_0, \alpha_1, \ldots, \alpha_{n-2}$ such that*

$$f \; = \; \sum_{i=0}^{n-2} \alpha_i \delta_{(a)}^{(i)}.$$

Proof: First note that for $n = 0$ or $n = 1$ the unique distribution which is null on $\mathbf{I}_1$ and on $\mathbf{I}_2$ is the distribution which is null on $\mathbb{R}$ (in the case that $f \in \mathcal{C}(\mathbb{R})$ the result is obvious; if $f = \mathbf{D}F$ with $F \in \mathcal{C}(\mathbb{R})$, from the conditions $f_{|\mathbf{I}_1} = \mathbf{0}$ and $f_{|\mathbf{I}_2} = \mathbf{0}$ it follows that F must be constant on $\mathbf{I}_1$ and on $\mathbf{I}_2$ and therefore on $\mathbb{R}$).

In the proof of the theorem we will suppose for the sake of simplicity that $a = 0$; we then pass to the more general case of an arbitrary $a \in \mathbb{R}$ by translation. Induction over n will be used throughout, starting from $n = 2$.

If $f \in \mathcal{C}_2(\mathbb{R})$, $f = \mathbf{D}^2 F$ with $F \in \mathcal{C}(\mathbb{R})$, the vanishing of f over the intervals $]-\infty, 0[$ and $]0, +\infty[$ implies the existence of constants $\alpha, \beta, \gamma \in \mathbf{C}$ such that

$$F(x) = \left\{ \begin{array}{ll} \alpha x + \beta & \text{if} \quad x \leq 0 \\ \gamma x + \beta & \text{if} \quad x > 0. \end{array} \right.$$

Setting $G(x) = F(x) - (\alpha x + \beta)$ we get $G(x) = (\gamma - \alpha)J(x)$ and therefore we obtain

$$f = \mathbf{D}^2 F = \mathbf{D}^2 G = \alpha_0 \delta,$$

where $\alpha_0 = \gamma - \alpha$.

Once having verified the case for $n = 2$ suppose now as induction hypothesis that the result holds for $n = p \geq 2$ and let $f \in \mathcal{C}_{p+1}(\mathbb{R})$ be a null distribution on $]-\infty, 0[$ and $]0, +\infty[$. Then if g is a primitive of f on $\mathbb{R}$, there will exist constants α', β' such that $g = \alpha'$ on $]-\infty, 0[$ and $g = \beta'$ on $]0, +\infty[$. Hence, setting

$$h = g - \alpha' - (\beta' - \alpha')\mathbf{H},$$

we will obtain $h \in \mathcal{C}_p(\mathbb{R})$ and $h = \mathbf{0}$ on $]-\infty, 0[$ and $]0, +\infty[$. By the induction hypothesis we may deduce that

$$h = \sum_{i=0}^{p-2} \alpha_i \delta^{(i)}$$

and therefore that

$$f = \mathbf{D}g = \mathbf{D}h + (\beta' - \alpha')\delta = \sum_{i=0}^{p-1} \beta_i \delta^{(i)}$$

with $\beta_0 = \beta' - \alpha'$ and, for $i > 0$, $\beta_i = \alpha_{i-1}$. $\qquad\qquad\square$

The preceding results may now be summarised as follows:

Theorem 1.15 *Let $a \in \mathbb{R}$, $\mathbf{I}_1 =]-\infty, a[$, $\mathbf{I}_2 =]a, +\infty[$, f_1 a distribution in $\mathcal{C}_\infty(\mathbf{I}_1)$ and f_2 a distribution in $\mathcal{C}_\infty(\mathbf{I}_2)$. For a common extension of f_1 and f_2 to $\mathbb{R}$ to exist it is necessary and sufficient that f_1 may be extended to $\bar{\mathbf{I}}_1$ and f_2 may be extended to $\bar{\mathbf{I}}_2$ (or, that both f_1 and f_2 may be extended to $\mathbb{R}$). When the condition is satisfied the extension is not unique but two extensions differ by a linear combination of derivatives of the distribution $\delta_{(a)}$.*

In the sequel and in the generality of the cases in which we will work with extensions of the kind considered in this theorem, we will be interested to ensure by additional appropriate conditions not only the existence but also the uniqueness of the extension. For such a supplementary condition to guarantee uniqueness it is enough, according to the final part of the statement of the theorem 1.15, that it cannot be satisfied by two different distributions which differ by a linear combination of derivatives of $\delta_{(a)}$.

Hence, for example, it would be enough to demand that the extensions should belong to a certain linear subspace of $\mathcal{C}_\infty(\mathbb{R})$, which did not contain any non-null linear combination of derivatives of $\delta_{(a)}$. Under such hypothesis, the difference between two common extensions of the distributions $f_1 \in \mathcal{C}(\mathbf{I}_1)$ and $f_2 \in \mathcal{C}(\mathbf{I}_2)$ - both belonging to the subspace considered - will be necessarily the null distribution, thus guaranteeing the uniqueness of the extension.

Keeping these ideas in mind several possibilities may arise, some of which lead naturally to the consideration of the concepts of value and limit of a distribution at a point. The next two sections will be dedicated to the study of these concepts.

This section will be terminated with a brief reference to the problem of the division of a distribution by functions of class $\mathcal{C}^\infty$. This problem may be stated in the following terms: given a distribution $g \in \mathcal{C}_\infty(\mathbb{R})$ and a function $\varphi \in \mathcal{C}^\infty(\mathbb{R})$ to determine $f \in \mathcal{C}_\infty(\mathbb{R})$ so that we have $\varphi f = g$.

If $\varphi(x) \neq 0$ for each $x \in \mathbb{R}$, it is obvious that the distribution $\frac{1}{\varphi}g$ is a solution and is actually the unique solution; for if f is any solution, then multiplying both members of the equality $\varphi f = g$ by $1/\varphi$ we obtain immediately $f = \frac{1}{\varphi}g$.

Excepting this trivial case, it is convenient to approach the general problem by starting with some simple examples. Consider first the equation:

$$x^m f(x) \;=\; 0, \tag{1.7}$$

with $m \in \mathbb{N}$. It is clear that any distribution which satisfies (1.7) will be null on $]-\infty, 0[$ and on $]0, +\infty[$, that is, will be a linear combination of derivatives of δ. It is therefore enough to take into account the formulas which refer to the products of the form $x^m \delta^{(n)}$ (example 1 in section 6) to conclude that the solution of (1.7) is constituted by the distributions of the form

$$\alpha_0 \delta(x) + \alpha_1 \delta'(x) + \cdots + \alpha_{m-1} \delta^{(m-1)}(x), \tag{1.8}$$

where $\alpha_0, \alpha_1, \ldots, \alpha_{m-1} \in \mathbf{C}$.

As a byproduct we may remark that the above formulas allow us to verify that for any $m, k \in \mathbb{N}$, the distribution

$$\theta(x) \;=\; (-1)^m \, \frac{k!}{(m+k)!} \, \delta^{(m+k)}(x)$$

satisfies the condition:

$$x^m \, \theta(x) \;=\; \delta^{(k)}(x);$$

it will be therefore easy (and soon useful) to obtain any solution for λ in an equation of the form:

$$x^m \, \lambda(x) \;=\; \nu(x),$$

where ν denotes a given distribution, null on $]-\infty, 0[$ and on $]0, +\infty[$.

Consider now the equation in f:

$$x^m \, f(x) \; = \; g(x), \tag{1.9}$$

where g is any distribution defined on $\mathbb{R}$.

Clearly if we determine one solution of this equation all the other solutions may be obtained by adding to this one a distribution of the form (1.8) (since the difference between any two solutions of (1.9) is certainly a solution of (1.7)).

To prove the existence of one solution set

$$h_1(x) \; = \; \frac{1}{x^m} \, g(x), \qquad \text{on } \mathbf{I}_1 = \,] - \infty, 0[$$

and

$$h_2(x) \; = \; \frac{1}{x^m} \, g(x), \qquad \text{on } \mathbf{I}_2 = \,]0, +\infty[.$$

Now proposition 1.13 shows that h_1 and h_2 are distributions which may be extended to $\mathbb{R}$ (since the restriction of g is clearly prolongable to each of the intervals $\mathbf{I}_1$ and $\mathbf{I}_2$). Hence, by theorem 1.15, h_1 and h_2 admit a common extension $h \in \mathcal{C}_\infty(\mathbb{R})$.

We will therefore have both on $\mathbf{I}_1$ and on $\mathbf{I}_2$:

$$x^m \, h(x) \; = \; x^m \left(\frac{1}{x^m} \, g(x) \right) \; = \; g(x)$$

and so the distribution

$$\nu(x) \; = \; x^m h(x) - g(x),$$

will be the null distribution on $\mathbf{I}_1$ and on $\mathbf{I}_2$.

Given a distribution λ such that

$$x^m \lambda(x) \; = \; \nu(x),$$

(which as already seen, is always possible) it is enough to put $f = h - \lambda$ to have

$$x^m f(x) \; = \; x^m h(x) - x^m \lambda(x) \; = \; g(x).$$

Accordingly we may now state

Theorem 1.16 *Any equation of the form*

$$x^m \, f(x) \; = \; g(x),$$

where $m \in \mathbb{N}$ and $g \in \mathcal{C}_\infty(\mathbb{R})$, has infinitely many solutions any two of which differ by a linear combination of derivatives of order $< m$ of the distribution δ.

As an example we remark that the general solution of the equation

$$x^m f(x) = 1$$

is

$$f(x) = \mathbf{Fp}\left\{\frac{1}{x^m}\right\} + \alpha_0\delta(x) + \alpha_1\delta'(x) + \cdots + \alpha_{m-1}\delta^{(m-1)}(x),$$

as the reader may easily verify.

In an analogous way we could prove the existence of solutions for any equation of the form

$$(x - a)^m f(x) = g(x),$$

and it would be confirmed that any two of those solutions would differ by a linear combination of $\delta_{(a)}, \ldots, \delta_{(a)}^{(m-1)}$.

Consider now the more general case of an equation of the form $\varphi f = g$ with given $\varphi \in \mathcal{C}^\infty(\mathbb{R})$ and $g \in \mathcal{C}_\infty(\mathbb{R})$ and suppose to start with that φ has one root of multiplicity m at a point[15] $a \in \mathbb{R}$ and does not vanish at any other point on an open interval $\mathbf{I}$ which contains the point a. It is then easy to see that, by procedures analogous to those just described, it is possible to determine those distributions $f \in \mathcal{C}_\infty(\mathbf{I})$ for which we have $\varphi f = g$ on the interval $\mathbf{I}$.

More generally, suppose that φ has roots of finite multiplicity at several points a_j (or even at infinitely many points provided they are isolated points, that is, so that in each bounded interval of $\mathbb{R}$ there are only a finite number of such points). It will then be possible to determine a family $\{\mathbf{I}_j\}$ of open intervals such that each interval $\mathbf{I}_j$ contains a unique point a_j and such that the union of all of them is the real line $\mathbb{R}$, and to solve the equation $\varphi f = g$ in each one of the intervals $\mathbf{I}_j$. Using the so-called *"piecewise joining principle"* (proved below in section 3.4) it can be shown that a distribution $f \in \mathcal{C}_\infty(\mathbb{R})$ is well defined if all its restrictions to the intervals $\mathbf{I}_j$ are known, where the union of the $\mathbf{I}_j$ gives $\mathbb{R}$. Then, under the above conditions, it would be possible to obtain from "local" solutions of the equation $\varphi f = g$ the "global" solutions (that is, the solutions of the equation defined on $\mathbb{R}$).

We observe finally that in the case that φ has roots of infinite degree of multiplicity the equation $\varphi f = g$ may have no solution. This is the case when, for example,

$$\varphi(x) = \mathrm{e}^{-1/|x|},$$

with $\varphi(0) = 0$ and $g(x) = 1$, as may be easily seen.

---

[15]That is, that φ is representable on a neighbourhood of a as $\varphi(x) = (x - a)^m\psi(x)$, with $\psi(a) \neq 0$ and ψ continuous at the point a.

Chapter 2
Value and Limit of a Distribution at a Point. Applications.

2.1   Value of a Distribution at a Point.

We have already observed that, in general, the notion of value of a distribution at any point of its domain is not defined. There are however important special cases in which it is possible to attribute, in a useful manner, a meaning to that concept.

The particular definition which will be considered in this section is basically due to the Polish mathematician Stefan Lojasiewicz.[1] The process that will be used here to introduce such a notion is however different from that used by Lojasiewicz and can be motivated by a brief reference to a method of summation of series due to Césaro.

In its most simple version, the Césaro method considers for a given series

$$\sum_{n=0}^{\infty} a_n$$

the arithmetical means:

$$\sigma_1 = s_1, \quad \sigma_2 = \frac{s_1 + s_2}{2}, \quad \ldots, \quad \sigma_n = \frac{s_1 + s_2 + \cdots + s_n}{n}, \quad \ldots$$

where $s_n = a_1 + a_2 + \cdots + a_n$.

It is not difficult to see that, if the series converges to the sum s (that is, the sequence s_n tends to the finite limit s) then $\lim_{n\to\infty} \sigma_n = s$ also holds. However, it may be that s_n, the sequence of the partial sums, has no finite limit but that the sequence of the arithmetical means σ_n does converge to a finite limit. In such a case, if $\lim_{n\to\infty} \sigma_n = \sigma$ it is usual to say that the series is convergent in the generalised sense (the so-called $(C,1)$-Césaro convergence) to the sum σ. For example the series

$$\sum_{n=0}^{\infty} (-1)^n,$$

which is divergent in the usual sense, converges in this generalised sense to the sum $\frac{1}{2}$.

A similar reasoning may be adopted to generalise the concept of the limit of a function at a point and, as we will see, to define the value and limit of a distribution at a point.

Consider, for example, a function F, continuous and bounded on an interval $]a, b]$, and denote by $\mu_a F(x)$ the mean value of F on the interval with end points a and x:

$$\mu_a F(x) \;=\; \frac{1}{x - a} \int_a^x F(t)\, dt \qquad x \in\,]a, b].$$

---

[1] A more general definition will be considered in the Appendix A.

It is immediately obvious that if F has a limit when $x \to a$, then $\mu_a F$ also has a limit at the same point and that

$$\lim_{x \to a} \mu_a F(x) \;=\; \lim_{x \to a} F(x).$$

But it may also happen that the limit of $\mu_a F$ does exist while the limit of F does not exist, and in this case we may say that F has a limit at the point a in a generalised sense. For example, the function $\sin \frac{1}{x}$ has no limit at the origin in the usual sense; however, since for any $x \neq 0$, we have

$$\frac{1}{x} \int_0^x \sin\left(\frac{1}{t}\right) dt \;=\; x \cos\left(\frac{1}{x}\right) - 2 \int_0^x \frac{t}{x} \cos\left(\frac{1}{t}\right) dt \tag{2.1}$$

where

$$\left| \frac{1}{x} \int_0^x \sin\left(\frac{1}{t}\right) dt \right| \;\leq\; 3|x|,$$

we can conclude that, in the generalised sense just referred to above, the limit of $\sin(\frac{1}{x})$ when $x \to 0$ exists and is zero.

The process may obviously be iterated: given the function F which we have supposed to be continuous and bounded on $]a, b]$, the function $G = \mu_a F$ is also continuous and bounded over the same interval. Moreover, if the limit $\lim_{x \to a} G(x)$ does not exist the "mean value" $\mu_a G$ may instead be considered; and so on successively.

We now start to formulate the concept of the value of a distribution at a point. Denote by $\mathbf{I}$ an interval of $\mathbb{R}$ and let a *be an interior point* of $\mathbf{I}$. If $F \in \mathcal{C}(\mathbf{I})$ then we denote by $\mu_a F$ the function continuous on I which is such that, for $x \in \mathbf{I} \backslash \{a\}$,

$$\mu_a F(x) \;=\; \frac{1}{x - a} \int_a^x F(t)\, dt.$$

We will clearly have $(\mu_a F)(a) = F(a)$.

Now denote by ∂_a the linear map from $\mathcal{C}_\infty(\mathbf{I})$ into $\mathcal{C}_\infty(\mathbf{I})$ that transforms each distribution g into the distribution

$$\partial_a g(\hat{x}) \;=\; \mathbf{D}\left[(\hat{x} - a) g(\hat{x})\right].$$

It is clear that, for each $F \in \mathcal{C}(\mathbf{I})$, we have $\partial_a (\mu_a F) = F$: the operator ∂_a (when restricted to the range of μ_a) is therefore a left inverse of μ_a. If, for each $p \in \mathbb{N}$ and each $f \in \mathcal{C}_\infty(\mathbf{I})$ we set as usual

$$\partial_a^0 f = f \quad \text{and} \quad \partial_a^{p+1} f = \partial_a\left(\partial_a^p f\right),$$

then we may now introduce the following definition:

Definition 2.1 *A distribution $f \in \mathcal{C}_\infty(\mathbf{I})$ is said to have a value at an interior point a of $\mathbf{I}$ if and only if there exist $p \in \mathbb{N}$ and $F \in \mathcal{C}(\mathbf{I})$ such that $f = \partial_a^p F$.*

Any distribution which has a value at a point $a \in \mathbf{I}$, in the sense of Definition 2.1, is said to be *continuous at a*.

We will denote by $V_a(\mathbf{I})$ (more simply by V_a) the subset of $\mathcal{C}_\infty = \mathcal{C}_\infty(\mathbf{I})$ comprising all distributions which are continuous at the point a. The following properties of V_a are immediate:

> i) $\mathcal{C} \subset V_a$, that is, the continuous functions on the interval $\mathbf{I}$ are continuous, as distributions, at the point a;
>
> ii) V_a is closed with respect to the operator ∂_a: $\partial_a(V_a) \subset V_a$.

It is easy to see that the map $\partial_a : V_a \to V_a$ is onto, that is, that $\partial_a(V_a) = V_a$: in fact, given a distribution $g \in V_a$, we have $g = \partial_a^n G$ with $G \in \mathcal{C}$ and then it is enough to put $f = \partial_a^n(\mu_a G)$ to have $\partial_a f = g$). It is also easy to see that *any subset of V_a which contains $\mathcal{C}$ and is closed with respect to the operator ∂_a necessarily coincides with V_a;* this property will be useful in the sequel.

We will now give some examples of distributions continuous at a certain point a (other than the elements of $\mathcal{C}$ themselves which are, of course, the first natural examples).

Let F be a function with indefinite integral on the interval $\mathbf{I}$ and a an interior point of $\mathbf{I}$ at which the function F has a finite limit (in the usual sense); under these conditions it is certain that $F \in V_a(\mathbf{I})$. To see this, it is enough to note that the function

$$\mu_a F(x) \;=\; \frac{1}{x-a} \int_a^x F(t)\, dt$$

(where $\mu_a F(a) = \lim_{x \to a} F(x)$) is an element of $\mathcal{C}(\mathbf{I})$ and to take into account properties i) and ii) referred to above.

Thus, for example, the function $\sin \frac{1}{x}$ (with indefinite integral on $\mathbb{R}$) is surely continuous, as a distribution, at any point $a \neq 0$. But it is easy to see that it is also continuous as a distribution at the point 0. In fact, denoting by θ the continuous function on $\mathbb{R}$ such that $\theta(x) = x \, \cos \frac{1}{x}$ for each $x \neq 0$, it may be confirmed from (2.1) that

$$\sin \frac{1}{\hat{x}} \;=\; \partial_0\left(\theta - 2\mu_0 \theta\right), \tag{2.2}$$

from which it follows at once that $\sin \frac{1}{x} \in V_0(\mathbb{R})$.

Before continuing it is convenient to remark that, if $f, g \in V_a(\mathbf{I})$ there always exist functions $F, G \in \mathcal{C}(\mathbf{I})$ and an integer n (the same in both cases) such that $f = \partial_a^n F$ and $g = \partial_a^n G$. In fact, the hypothesis $f, g \in V_a$ ensures the existence of $p, q \in \mathbb{N}$ and $\Phi, \Psi \in \mathcal{C}$ such that $f = \partial_a^p \Phi$, $g = \partial_a^q \Psi$, and now it is enough to set $n = \max\{p, q\}$, $F \in \mu_a^{n-p}\Phi$ and $G = \mu_a^{n-q}\Psi$.

We may now state

Proposition 2.2 $V_a(\mathbf{I})$ *is a linear subspace of $\mathcal{C}_\infty(\mathbf{I})$.*

Proof: Let $\alpha, \beta \in \mathbf{C}$ and $f, g \in V_a$; if we suppose that $f = \partial_a^n F$ and $g = \partial_a^n G$, we will have

$$\alpha f + \beta g = \partial_a^n (\alpha F + \beta G) \in V_a,$$

which proves that V_a is a subspace of $\mathcal{C}_\infty$. $\qquad\qquad \square$

Another result of great interest in the sequel is the following

Theorem 2.3 *The restriction of ∂_a to the space $V_a(\mathbf{I})$ is an automorphism on this linear space.*

Proof: Denote again by ∂_a the restriction of the operator ∂_a defined above to the space V_a. Clearly ∂_a is a linear map from V_a into itself which, as already seen, is also an onto map. To prove that it is an injection, (which will complete the proof of the theorem) it will be enough to verify that, if $\partial_a f = 0$ with $f \in V_a$ then $f = 0$; for that it will obviously be sufficient to show that the equality $\partial_a^p F = 0$, with $F \in \mathcal{C}$ and $p \in \mathbb{N}$, can only be satisfied if $F = 0$.

Suppose then that $\partial_a^p F = 0$, set $\mathbf{I}_1 = \mathbf{I} \cap] -\infty, a[$ and denote by F_1 the restriction of F to the interval $\mathbf{I}_1$. F_1 is a continuous function on $\mathbf{I}_1$, with finite limit when $x \to a$.

If $\Phi(x)$ is a continuous function on $\mathbf{I}_1$, then the change of variable $x = a - \mathrm{e}^t$ transforms the distribution $\partial_a^p \Phi$ into a distribution, defined on a given interval $\mathbf{J}$ unbounded to the left, and of the form $\mathrm{e}^{-t} \mathbf{D}_t^p \Psi(t)$, with $\Psi(t) = \mathrm{e}^t \Phi(a - \mathrm{e}^t)$. It follows immediately that, by applying this change of variable to each member of the equality $\partial_a^p F_1 = 0$, we must have $\mathbf{D}_t^p G_1(t) = 0$ in $\mathbf{J}$, with

$$G_1(t) = \mathrm{e}^t F_1(a - \mathrm{e}^t).$$

$G_1(t)$ will be therefore a polynomial in t of degree less than p and, since F_1 has a finite limit when $x \to a$, it must be the case that $\lim_{t \to -\infty} G_1(t) = 0$. This result implies in its turn that $G_1(t) = 0$ and therefore that $F_1 = 0$.

Similarly it can be verified that the restriction of F to the interval $\mathbf{I}_2 = \mathbf{I} \cap]a, +\infty[$ is the null function. Therefore, taking into account that $F \in \mathcal{C}(\mathbf{I})$ we get $F = 0$. $\qquad \square$

Proposition 2.4 *If a is an interior point of the interval $\mathbf{I}$, the Dirac distribution located at the point a, $\delta_{(a)}$, does not belong to the space $V_a(\mathbf{I})$.*

Proof: From the injectivity of the mapping $\partial_a : V_a(\mathbf{I}) \to V_a(\mathbf{I})$ it follows that we can only have $\partial_a g = 0$ with $g \in V_a(\mathbf{I})$ if we have $g = 0$; but from the equality $(x - a)\delta(x - a) = 0$ it follows at once, by differentiation, that $\partial_a \left(\delta_{(a)} \right) = 0$. $\qquad \square$

Before we state the theorem that follows we may remark that, also as a consequence of the injectivity of $\partial_a : V_a(\mathbf{I}) \to V_a(\mathbf{I})$, an equality of the form $\partial_a^n f = \partial_a^n g$ with $f, g \in V_a(\mathbf{I})$ necessarily implies that $f = g$.

Theorem 2.5 *There exists one and only one map $f \rightsquigarrow f(a)$ from $V_a(\mathbf{I})$ into $\mathbf{C}$ satisfying the following two conditions:*

i) *if $f \in \mathcal{C}(\mathbf{I})$, $f(a)$ is the usual value of f at the point a;*

ii) *for any $f \in V_a(\mathbf{I})$, $\partial_a f(a) = f(a)$.*

Proof: The uniqueness is immediate: if f is an element in V_a, and therefore representable in the form $f = \partial_a^p F$, with $p \in \mathbb{N}$ and $F \in \mathcal{C}$, then the conditions **i)** and **ii)** in the theorem ensure that when the mapping does in fact exist its value will be uniquely determined: $f(a)$ cannot be different from $F(a)$.

To prove existence note that, if $f = \partial_a^p F = \partial_a^q G$, then setting $r = \max\{p, q\}$ we will have

$$\partial_a^r \left(\mu_a^{r-p} F \right) = \partial_a^r \left(\mu_a^{r-q} G \right)$$

and therefore

$$\mu_a^{r-p} F = \mu_a^{r-q} G$$

which implies that

$$F(a) = \mu_a^{r-p} F(a) = \mu_a^{r-q} G(a) = G(a).$$

It is therefore proved that we effectively define a mapping from V_a into $\mathbb{C}$ if we associate to each $f = \partial_a^p F \in V_a$ the complex number $f(a) = F(a)$. Moreover, since the mapping $f \rightsquigarrow f(a)$ defined in this way satisfies **i)** and **ii)**, then the proof is complete. $\square$

By definition, the number $f(a)$ associated with the distribution $f \in V_a$ as stated in the above theorem, is called *the value of the distribution f at the point a*. Thus, for example, from the equality (2.2) it follows at once that the distribution $\sin \frac{1}{x}$ has the value 0 at the origin.

We will have the opportunity to verify in the sequel that the concept just defined of the value of a distribution at a point satisfies many desirable properties, in particular those relating to the fundamental operations on the distributions defined in preceeding sections.

Proposition 2.6 *The map $f \rightsquigarrow f(a)$ from V_a into $\mathbb{C}$ is linear.*

Proof: Let $\alpha, \beta \in \mathbb{C}$ and $f, g \in V_a$; representing these distributions in the form $f = \partial_a^n F$, $g = \partial_a^n G$, we will have $\alpha f + \beta g = \partial_a^n (\alpha F + \beta G)$ and therefore:

$$(\alpha f + \beta g)(a) = (\alpha F + \beta G)(a) = \alpha F(a) + \beta G(a) = \alpha f(a) + \beta g(a)$$

as was to be proved. $\square$

Proposition 2.7 *If $\varphi \in \mathcal{C}^\infty(\mathbf{I})$ and $f \in V_a(\mathbf{I})$ then $\varphi f \in V_a(\mathbf{I})$ and moreover, $(\varphi f)(a) = \varphi(a) f(a)$.*

Proof: Denote by E_a the subset of V_a comprising all distributions f which are such that

$$\varphi f \in V_a \quad \text{and} \quad (\varphi f)(a) = \varphi(a) f(a),$$

for every function $\varphi \in \mathcal{C}^\infty$. The proposition will be proved if we can show that $E_a = V_a$ and, for that, taking into account the obvious relation $\mathcal{C} \subset E_a$, it will be enough to show that, if $f \in E_a$, then $\partial_a f \in E_a$ also holds.

Let $f \in E_a$. Setting $g = \partial_a f$, consider the equality, valid for any $\varphi \in \mathcal{C}^\infty$:

$$\varphi g = \varphi \partial_a f = \partial_a(\varphi f) - \psi f,$$

where $\psi(x) = (x - a)\varphi'(x)$ for each $x \in \mathbf{I}$. Since $f \in E_a$ we have, on the one hand, $\varphi f \in V_a$ and $\psi f \in V_a$, and on the other hand $(\varphi f)(a) = \varphi(a)f(a)$ and $(\psi f)(a) = \psi(a)f(a) = 0$. The preceding equality therefore allows us to see that we must have $\varphi g \in V_a$ and

$$
\begin{aligned}
(\varphi g)(a) &= [\partial_a(\varphi f)](a) - (\psi f)(a) \\
&= (\varphi f)(a) = \varphi(a)f(a) = \varphi(a)g(a).
\end{aligned}
$$

It may therefore be concluded that $g = \partial_a f \in E_a$, and this ends the proof. $\square$

Proposition 2.8 *Let* $\mathbf{I}$ *and* $\mathbf{J}$ *be two intervals of* $\mathbb{R}$, $\varphi : \mathbf{J} \to \mathbf{I}$ *a function in* C^∞ *such that* $\varphi'(t) \neq 0$ *for each* $t \in \mathbf{J}$, c *an interior point in* $\mathbf{J}$ *and* $a = \varphi(c)$;[2] *suppose also that* $f \in \mathcal{C}_\infty(\mathbf{I})$.

Under these conditions, if $f \in V_a(\mathbf{I})$, *then* $f \circ \varphi \in V_c(\mathbf{J})$ *and we have the equality* $f(a) = (f \circ \varphi)(c)$.

Proof: Denote by F_a the set of distributions $f \in V_a(\mathbf{I})$ for which we have $f \circ \varphi \in V_c(\mathbf{J})$ and $(f \circ \varphi)(c) = f(a)$.

Since $\mathcal{C} \subset F_a$, the proposition will be proved provided we can show that we have $\partial_a f \in F_a$ whenever $f \in F_a$.

Accordingly let $f \in F_a$. By setting $g = \partial_a f$ and applying to both members of this equality the change of variable $x = \varphi(t)$ we obtain:

$$
(g \circ \varphi)(t) = \frac{1}{\varphi'(t)} \, \mathbf{D}_t \left[(\varphi(t) - \varphi(c))(f \circ \varphi)(t) \right].
$$

If we denote by θ the function defined on $\mathbf{J}$ such that $\theta(c) = \varphi'(c)$ and

$$
\theta(t) = \frac{\varphi(t) - \varphi(c)}{t - c} \quad \text{if} \ \ t \in \mathbf{J} \setminus \{c\}
$$

then we will have[3] $\theta \in C^\infty(\mathbf{J})$ and

$$
(g \circ \varphi)(t) = \frac{1}{\varphi'(t)} \, \mathbf{D}_t \left[(t - c)\theta(t)(f \circ \varphi)(t) \right],
$$

or even

$$
(g \circ \varphi)(t) = \frac{1}{\varphi'(t)} \, \partial_c^* \left[\theta(t)(f \circ \varphi)(t) \right],
$$

(where $\partial_c^* h(t) = \mathbf{D}_t[(t - c)h(t)]$, for each $h \in \mathcal{C}_\infty(\mathbf{J})$).

Taking into account the definition of the set F_a, the hypothesis $f \in F_a$ and the proposition 2.7, it may be concluded that $g \circ \varphi \in V_c(\mathbf{J})$ and that

$$
(g \circ \varphi)(c) = \frac{1}{\varphi'(c)} \, [\theta(c)(f \circ \varphi)(c)] = f(a) = g(a),
$$

and this completes the proof. $\square$

---

[2] Under the conditions of the hypothesis it is clear that a will be an interior point of $\mathbf{I}$.

[3] To see this it is enough to make the change of variable $r = c + (t - c)s$ in the integral of the right hand side of the equality

$$
\varphi(t) - \varphi(c) = \int_c^t \varphi'(r)\,dr.
$$

Proposition 2.9 *Let a be an interior point of $\mathbf{I}$ and $f \in \mathcal{C}_\infty(\mathbf{I})$; if we have* $\mathbf{D}f \in V_a(\mathbf{I})$ *then* $f \in V_a(\mathbf{I})$ *also holds and there exists* $h \in V_a(\mathbf{I})$ *such that*

$$f(x) - f(a) = (x - a)h(x)$$

and $h(a) = \mathbf{D}f(a)$.

Proof: Denote by h the (unique) distribution belonging to V_a that satisfies the condition $\partial_a h = \mathbf{D}f$. We will have $h(a) = \mathbf{D}f(a)$ and, from the equality

$$\mathbf{D}f(x) = \mathbf{D}[(x - a)h(x)]$$

it can be said that, for some $c \in \mathbf{C}$,

$$f(x) = (x - a)h(x) + c.$$

Since the distribution on the right-hand side belongs to the space V_a, we also have $f \in V_a$ and (replacing x by a in the last equality), $c = f(a)$. That is,

$$f(x) - f(a) = (x - a)h(x)$$

as was to be proved. $\qquad\qquad\square$

If a is an interior point of the interval $\mathbf{I}$, then the proposition 2.9 and the fact that $\delta_{(a)} \notin V_a(\mathbf{I})$ show that, for any $p \in \mathbb{N}$, $\delta_{(a)}^{(p)} \notin V_a(\mathbf{I})$ also holds. But more generally it is easily seen that no non-null linear combination of derivatives of the distribution $\delta_{(a)}$ belongs to the space $V_a(\mathbf{I})$; in fact, if g is a distribution of the form:

$$g(x) = \sum_{k=0}^{m} c_k \delta^{(k)}(x - a),$$

with $m \in \mathbb{N}$, $c_k \in \mathbf{C}$ and $c_m \neq 0$, we have (see sections 1.6 and 1.7):

$$(x - a)^m g(x) = (-1)^m m! c_m \delta(x - a).$$

Now, since $\delta_{(a)} \notin V_a(\mathbf{I})$, it follows from proposition 2.7 that $g \notin V_a(\mathbf{I})$.

Taking proposition 1.13 into account it can now be said that there exists only one distribution, defined on $\mathbf{I}$, which is null on $\mathbf{I}_1 = \mathbf{I} \cap]-\infty, a[$ and on $\mathbf{I}_2 = \mathbf{I} \cap]a, +\infty[$ and is continuous at the point a: this is the null distribution itself. From this it follows in particular that the unique distribution $f \in V_a(\mathbf{I})$ that satisfies an equality of the form

$$(x - a)^p f(x) = 0,$$

with $p \in \mathbb{N}$, is precisely the null distribution.

Proposition 2.9 has an important extension, which may be expressed as an equality which it is natural to refer to as *"the Taylor formula"*. Before stating this it is convenient to state two formulas, (2.3) and (2.4), which will be useful in this context. Proof of these is left to the reader as a simple exercise.

Let $\mathbf{I}$ be an interval of $\mathbb{R}$, a an arbitrary point of $\mathbb{R}$, $f \in \mathcal{C}_\infty(\mathbf{I})$, $n, p \in \mathbb{N}$; then

$$\partial_a^n \left[(x-a)^p f(x)\right] \;=\; (x-a)^p \sum_{k=0}^{n} \binom{n}{k} p^k \partial_a^{n-k}[f(x)] \tag{2.3}$$

$$(x-a)^p \partial_a^n [f(x)] \;=\; \sum_{k=0}^{n} (-1)^k \binom{n}{k} p^k \partial_a^{n-k} \left[(x-a)^p f(x)\right]. \tag{2.4}$$

With regard to (2.3) it should be noted that, when a is an interior point of $\mathbf{I}$ and f is continuous at a, the sum on the right-hand side also denotes a distribution which is continuous at the same point. Hence it can be said that, if $f \in V_a(\mathbf{I})$ and $n, p \in \mathbb{N}$, there always exists $g \in V_a(\mathbf{I})$ such that

$$\partial_a^n \left[(x-a)^p f(x)\right] \;=\; (x-a)^p g(x). \tag{2.5}$$

But it is easy to verify that there is only one distribution g satisfying the conditions referred to above; in fact, if we also have

$$\partial_a^n \left[(x-a)^p f(x)\right] \;=\; (x-a)^p g^*(x),$$

with $g^* \in V_a(\mathbf{I})$, then

$$(x-a)^p \left[g(x) - g^*(x)\right] \;=\; 0$$

and, since $g - g^*$ is continuous at the point a, this implies that $g^* = g$.

We may therefore adopt the following convention: if $f \in V_a(\mathbf{I})$ and $n, p \in \mathbb{N}$, we will denote by $(x-a)^{-p} \partial_a^n[(x-a)^p f(x)]$ the unique distribution $g \in V_a(\mathbf{I})$ which satisfies (2.5). Note that it follows immediately from (2.3), that if $f(a) = 0$ then we also have $g(a) = 0$.

We may now state the following lemma:

Lemma 2.10 *Let a be an interior point of the interval* $\mathbf{I}$, *let $p \in \mathbb{N}$ and f belong to $\mathcal{C}_\infty(\mathbf{I})$. If there exists a distribution $g \in V_a(\mathbf{I})$ such that*

$$\mathbf{D}f(x) \;=\; \frac{(x-a)^p}{p!}\, g(x)$$

then there also exists $h \in V_a(\mathbf{I})$ such that

$$f(x) - f(a) \;=\; \frac{(x-a)^{p+1}}{(p+1)!}\, h(x)$$

and, moreover, $h(a) = g(a)$.

Proof: In the first place consider the case in which $g(a) = 0$. Under the hypothesis there exist $n \in \mathbb{N}$ and $G \in \mathcal{C}(\mathbf{I})$ such that

$$\mathbf{D}f(x) \;=\; \frac{(x-a)^p}{p!}\, \partial_a^n G(x) \;=\; \sum_{k=0}^{n} c_k \partial_a^{n-k} \left[(x-a)^p G(x)\right], \tag{2.6}$$

with $c_k = (-1)^k \binom{n}{k} p^k / p!$ and $G(a) = g(a) = 0$.

However, from the equality:

$$\mathbf{D}f(x) = \sum_{k=0}^{n-1} c_k \mathbf{D}\left\{(x-a)\partial_a^{n-k-1}\left[(x-a)^p G(x)\right]\right\} + c_n(x-a)^p G(x),$$

which is equivalent to (2.6), and from the fact that, by hypothesis, f must be continuous at a, it follows immediately by antiderivation that:

$$f(x) - f(a) = \sum_{k=0}^{n-1} c_k(x-a)\partial_a^{n-k-1}\left[(x-a)^p G(x)\right] + c_n(x-a)\mu_a\left[(x-a)^p G(x)\right].$$

Under these conditions, set

$$h(x) = \sum_{k=0}^{n-1} c_k(x-a)^{-p}\partial_a^{n-k-1}\left[(x-a)^p G(x)\right] + c_n\theta(x),$$

where θ denotes a continuous function on $\mathbf{I}$ which is such that, for each $x \in \mathbf{I}\backslash\{a\}$,

$$\theta(x) = \frac{1}{(x-a)^{p+1}} \int_a^x (t-a)^p G(t)dt.$$

Then it can easily be seen, on the one hand that

$$\frac{(x-a)^{p+1}}{(p+1)!} h(x) = f(x) - f(a)$$

and on the other hand, that the equality $h(a) = 0$ holds (taking into account that all distributions

$$(x-a)^{-p}\partial_a^{n-k-1}\left[(x-a)^p G(x)\right]$$

and θ have the value 0 at the point a).

Now, in the more general setting, suppose that we have

$$\mathbf{D}f(x) = \frac{(x-a)^p}{p!} g(x),$$

with $g(a) = \alpha$. Setting

$$f^*(x) = f(x) - \alpha \frac{(x-a)^{p+1}}{(p+1)!}$$

we will have

$$\mathbf{D}f^*(x) = \mathbf{D}f(x) - \alpha \frac{(x-a)^p}{p!} = \frac{(x-a)^p}{p!} g^*(x),$$

with $g^*(a) = g(a) - \alpha = 0$.

We must therefore have

$$f^*(x) - f^*(a) = \frac{(x-a)^{p+1}}{(p+1)!} h^*(x),$$

with $h^*(0) = 0$ and, since

$$f^*(x) - f^*(a) = f(x) - \alpha \frac{(x-a)^{p+1}}{(p+1)!} - f(a),$$

it follows that

$$f(x) - f(a) = \frac{(x-a)^{p+1}}{(p+1)!}\, h(x),$$

where $h(x) = h^*(x) + \alpha$ is a distribution which is continuous at the point a and such that $h(a) = \alpha$. $\qquad\square$

We may now state the theorem:

Theorem 2.11 *Let a be an interior point of the interval $\mathbf{I}$ and let f be a distribution defined on $\mathbf{I}$ such that $\mathbf{D}^n f \in V_a(\mathbf{I})$. Then there exists h in $V_a(\mathbf{I})$ such that $h(a) = \mathbf{D}^n f(a)$ and*

$$\begin{aligned} f(x) \;=\;& f(a) + (x-a)\mathbf{D}f(a) + \dots \\ &+ \frac{(x-a)^{n-1}}{(n-1)!}\,\mathbf{D}^{n-1}f(a) + \frac{(x-a)^n}{n!}\,h(x). \end{aligned}$$

Proof: The proof will be by induction on n. For $n = 1$, the result coincides with that of Proposition 2.9. Suppose now that it is true for $n = p$ and let f be a distribution defined on $\mathbf{I}$ and such that $\mathbf{D}^{p+1} f \in V_a(\mathbf{I})$.

The induction hypothesis (applied to $\mathbf{D}f$) guarantees that there exists a distribution g such that

$$\mathbf{D}f(x) - \left[\mathbf{D}f(a) + (x-a)\mathbf{D}^2 f(a) + \cdots + \frac{(x-a)^{p-1}}{(p-1)!}\,\mathbf{D}^p f(a)\right] = \frac{(x-a)^p}{p!}\,g(x),$$

with $g(a) = \mathbf{D}^{p+1} f(a)$. Taking into account lemma 2.10 it can therefore be said that there exists a distribution h which satisfies the following conditions:

$$f(x) - \left[f(a) + (x-a)\mathbf{D}f(a) + \cdots + \frac{(x-a)^p}{p!}\,\mathbf{D}^p f(a)\right] = \frac{(x-a)^{p+1}}{(p+1)!}\,h(x)$$

where $h(a) = g(a) = \mathbf{D}^{p+1} f(a)$. This completes the proof. $\qquad\square$

This section will be ended with a reference to what may be called the local character of the concept of the value of a distribution at a point.

Suppose that f and g are two distributions defined on the interval $\mathbf{I}$ and that g is continuous at a certain point a; further, suppose that there exists an interval $\mathbf{I}' \subset \mathbf{I}$, that a is an interior point of $\mathbf{I}'$ and that $f = g$ on $\mathbf{I}'$. Will it be possible to guarantee, under these conditions, that f is continuous at the point a and that $f(a) = g(a)$?

To show that the answer to this question is the affirmative we begin by proving the following proposition:

Proposition 2.12 *Let $\mathbf{I}$ and $\mathbf{I}'$ be two intervals such that $\mathbf{I}' \subset \mathbf{I}$, let a be an interior point of $\mathbf{I}'$ and let f be a distribution defined on $\mathbf{I}$; then $f \in V_a(\mathbf{I})$ if and only if $f_{|\mathbf{I}'} \in V_a(\mathbf{I}')$ and, if this condition is satisfied, we also have $f(a) = f_{|\mathbf{I}'}(a)$.*

Proof: Since the equality $f = \partial_a^n F$ (with $n \in \mathbb{N}$, $F \in \mathcal{C}(\mathbf{I})$) implies $f_{|\mathbf{I}'} = \partial_a^n F_{|\mathbf{I}'}$, it is clear that, if we have $f \in V_a(\mathbf{I})$ then $f_{|\mathbf{I}'} \in V_a(\mathbf{I}')$ and $f_{|\mathbf{I}'}(a) = f(a)$; hence, it only remains to prove that the conditions $f \in \mathcal{C}_\infty(\mathbf{I})$ and $f_{|\mathbf{I}'} \in V_a(\mathbf{I}')$ imply that $f \in V_a(\mathbf{I})$.

The proof follows along similar lines to that of theorem 2.3, using change of variables. To simplify notation we prove for the case $a = 0$ (to which we can always reduce the general problem by a simple translation); also we write ∂ for ∂_0 throughout.

Suppose then that the origin is an interior point of the interval $\mathbf{I}'$, that $\mathbf{I}' \subset \mathbf{I}$ and that there exist $n \in \mathbb{N}$, $F \in \mathcal{C}(\mathbf{I}')$ and $G \in \mathcal{C}(\mathbf{I})$ such that $f = \partial^n F$ on $\mathbf{I}'$ and $f \in \mathbf{D}^n G$ on $\mathbf{I}$. Set

$$\begin{aligned}
\mathbf{I}_2 &= \mathbf{I} \cap \,]0, +\infty[\,, & \mathbf{I}_2' &= \mathbf{I}' \cap \,]0, +\infty[\,, \\
\mathbf{J} &= \{\log x : x \in \mathbf{I}_2\}\,, & \mathbf{J}' &= \{\log x : x \in \mathbf{I}_2'\}\,,
\end{aligned}$$

and denote by φ the bijection from $\mathbf{J}$ into $\mathbf{I}_2$ defined by $x = \varphi(t) = \exp(t)$, $t \in \mathbf{J}$ and by ψ the restriction to $\mathbf{J}'$ of the function φ.

Under these conditions, the change of variable $x = \psi(t)$ transforms $f_{|\mathbf{I}_2'}$ into a distribution $g \in \mathcal{C}_\infty(\mathbf{J}')$ such that

$$g(t) = \mathrm{e}^{-t} \mathbf{D}_t^n \left[\mathrm{e}^t F\left(\mathrm{e}^t\right) \right] \quad (\text{on } \mathbf{J}'), \tag{2.7}$$

while the change of variable $x = \varphi(t)$ transforms $f_{|\mathbf{I}_2}$ into a distribution defined on $\mathbf{J}$, which we will denote by $h(t)$.

Since the degree of the distribution $h(t)$ is certainly $\leq n$, the product $\exp(t)h(t)$ may be represented in the form

$$\mathrm{e}^t h(t) = \mathbf{D}_t^n K(t), \tag{2.8}$$

with $K \in \mathcal{C}(\mathbf{J})$.

However, since clearly we have $g = h_{|\mathbf{J}'}$, it follows from (2.7) and (2.8) that a polynomial $P(t)$ exists such that

$$\mathrm{e}^t F\left(\mathrm{e}^t\right) = K(t) + P(t) \quad (\text{on } \mathbf{J}').$$

Hence, if we set, for each $t \in \mathbf{J}$:

$$\Phi(t) = \mathrm{e}^{-t}[K(t) + P(t)]$$

we will have:

$$\Phi(t) = F\left(\mathrm{e}^t\right) \quad (\text{on } \mathbf{J}') \tag{2.9}$$

and

$$h(t) = \mathrm{e}^{-t} \mathbf{D}_t^n \left[\mathrm{e}^t \Phi(t) \right] \quad (\text{on } \mathbf{J}).$$

By the inverse change of variable $x = \varphi(t)$ we obtain on $\mathbf{I}_2$:

$$f(x) = \partial^n \Phi(\log x).$$

Taking (2.9) into account it is easily seen that there exists a function $\tilde{F}$, continuous on the interval $\mathbf{I}' \cup \mathbf{I}_2$ and such that:

$$\begin{aligned}
\tilde{F}(x) &= F(x) & (\text{on } \mathbf{I}') \\
\tilde{F}(x) &= \Phi(\log x) & (\text{on } \mathbf{I}_2)
\end{aligned}$$

and $f = \partial^n \tilde{F}$ (on $\mathbf{I}' \cup \mathbf{I}_2$).

It is clear that by a process analogous to that used to show that the function $F \in \mathcal{C}(\mathbf{I}')$ may be extended to a function $\tilde{F} \in \mathcal{C}(\mathbf{I}' \cup \mathbf{I}_2)$, such that $f = \partial^n \tilde{F}$ (on $\mathbf{I}' \cup \mathbf{I}_2$), we could prove that the function $\tilde{F}$ in its turn may be extended into a function $F^* \in \mathcal{C}(\mathbf{I})$, such that $f = \partial^n F^*$ (on $\mathbf{I}$). We may therefore consider the proof complete. $\qquad \square$

We may now state the following corollary:

> **Corollary 2.13** *Let $f, g \in \mathcal{C}_\infty(\mathbf{I})$, $\mathbf{I}'$ an interval contained in $\mathbf{I}$ and a an interior point of $\mathbf{I}'$. If $f = g$ on $\mathbf{I}'$ and if $g \in V_a(\mathbf{I})$, then $f \in V_a(\mathbf{I})$ and $f(a) = g(a)$.*

Proof: From $g \in V_a(\mathbf{I})$ it follows that $f_{|\mathbf{I}'} = g_{|\mathbf{I}'} \in V_a(\mathbf{I}')$ and $g(a) = g_{|\mathbf{I}'}(a) = f_{|\mathbf{I}'}(a)$. The proposition 2.12 therefore implies that $f \in V_a(\mathbf{I})$ and that $f(a) = f_{|\mathbf{I}'}(a) = g(a)$. $\qquad \square$

2.2   Limit of a Distribution at Infinity and at a Point $a \in \mathbb{R}$.

Again, consider the following problem: if $a \in \mathbb{R}$, $\mathbf{I}_1 =]-\infty, a[$ and $\mathbf{I}_2 =]a, +\infty[$ then we want to determine a common extension to $\mathbb{R}$ of two given distributions $f_1 \in \mathcal{C}_\infty(\mathbf{I}_1)$ and $f_2 \in \mathcal{C}_\infty(\mathbf{I}_2)$.

With only these conditions we already know that the problem is only possible if the distributions f_1 and f_2 may simultaneously extend into $\mathbb{R}$, the extension being indeterminate in this case. We have already seen that if we impose the additional condition that the extension of f be continuous at the point a, the problem cannot have more than one solution. But it is clear that the condition that f_1 and f_2 may extend into $\mathbb{R}$ will not be sufficient to ensure the existence of a (continuous) extension.

With regard to this point it is convenient to recall that in classical Analysis the condition for a function F, defined on $\mathbb{R} \backslash \{a\}$, to extend by continuity to the point a is that the limit of $F(x)$ when $x \to a$ should exist. We will be able to see that, in a distributional setting, an entirely analogous condition is valid. To show this it will be necessary to introduce the concept of the limit of a distribution at a point $a \in \mathbb{R}$. It is actually simpler, and more convenient, to treat first the notion of the limit of a distribution at infinity.

Suppose that $\mathbf{I}$ is an interval of the form $]a, +\infty[$, where $a \geq 0$. Given an arbitrarily fixed point $c \in \mathbf{I}$, for each $F \in \mathcal{C}(\mathbf{I})$ denote by $\tilde{\mu}_c F$ (or simply $\tilde{\mu} F$) the function defined by:

$$\tilde{\mu} F(x) = \frac{1}{x} \int_c^x F(t) dt \qquad (x \in \mathbf{I}).$$

In addition, let us denote by $\mathcal{C}(\mathbf{I}; +\infty)$ the subset of $\mathcal{C}(\mathbf{I})$ comprising the functions F for which the limit of $F(x)$ when $x \to +\infty$ exists and is finite.

Clearly, if $F \in \mathcal{C}(\mathbf{I}; +\infty)$ then we also have that $\tilde{\mu} F \in \mathcal{C}(\mathbf{I}; +\infty)$, and the equality

$$\lim_{x \to +\infty} \tilde{\mu} F(x) = \lim_{x \to +\infty} F(x)$$

holds.

Still denoting by ∂ the mapping from $\mathcal{C}_\infty(\mathbf{I})$ into itself such that

$$\partial f = \mathbf{D}\left[\hat{x} f(\hat{x})\right] \qquad (f \in \mathcal{C}_\infty(\mathbf{I}))$$

we will have $\partial \tilde{\mu} F = F$ for every $F \in \mathcal{C}(\mathbf{I})$.

We will now introduce the following definition: the distribution $f \in \mathcal{C}_\infty(\mathbf{I})$ will be said to be *convergent at the point* $+\infty$, or *have a limit when* $x \to +\infty$ if and only if there exist $n \in \mathbb{N}$ and $F \in \mathcal{C}(\mathbf{I}; +\infty)$ such that $f = \partial^n F$. If, for example, Φ is a function with indefinite integral on $\mathbf{I}$ and with finite limit (in the usual sense) when $x \to +\infty$, Φ will be identified with the distribution:

$$\mathbf{D}\int_c^x \Phi(t)dt = \partial\left[\frac{1}{x}\int_c^x \Phi(t)dt\right]$$

and therefore it will be convergent at $+\infty$ in the distributional sense.

In the sequel we will denote by $\Lambda_{+\infty}(\mathbf{I})$ (or, more simply, by $\Lambda_{+\infty}$) the set of all distributions which are defined on $\mathbf{I}$ and convergent at the point $+\infty$.

Clearly, if $f, g \in \Lambda_{+\infty}$ there always exist functions $F, G \in \mathcal{C}(\mathbf{I}; +\infty)$ and an integer $n \geq 0$ such that $f = \partial^n F$ and $g = \partial^n G$. From this it follows immediately that $\Lambda_{+\infty}(\mathbf{I})$ is a linear subspace of $\mathcal{C}_\infty(\mathbf{I})$. It is also easily seen that the restriction of ∂ to $\Lambda_{+\infty}$ is a linear map and that $\partial(\Lambda_{+\infty}) = \Lambda_{+\infty}$; however, considered as a mapping from $\Lambda_{+\infty}$ into itself, ∂ is not an injection. In fact, the condition

$$\partial f(x) = \mathbf{D}[x f(x)] = 0$$

is satisfied if we have

$$f(x) = \frac{c_1}{x},$$

for some constant $c_1 \in \mathbf{C}$ and it is obvious that any distribution of this form is an element of $\mathcal{C}(\mathbf{I}; +\infty) \subset \Lambda_{+\infty}(\mathbf{I})$ (since we have supposed from the very beginning that $\mathbf{I} \subset \,]0, +\infty[$).

More generally, by induction, one can see that, for every $n \in \mathbb{N}_1$, we have $\partial^n f = 0$, with $f \in \mathcal{C}_\infty(\mathbf{I})$ if and only if:

$$f(x) = \frac{1}{x}\left[c_1 + c_2 \log x + \cdots + c_n \log^{n-1} x\right],$$

with $c_1, c_2, \ldots, c_n \in \mathbf{C}$. In this sense, all distributions f that satisfy the condition $\partial^n f = 0$ are continuous functions on $\mathbf{I}$, with limit zero when $x \to +\infty$.

The following theorem may now be proved:

Theorem 2.14 *There exists one and only one mapping* $f \rightsquigarrow f(+\infty)$, *(equal to* $\lim_{+\infty} f = \lim_{x \to +\infty} f(x)$*) from* $\Lambda_{+\infty}(\mathbf{I})$ *into* $\mathbf{C}$ *satisfying the following conditions:*

i) *if* $f \in \mathcal{C}(\mathbf{I}; +\infty)$, $f(+\infty)$ *is the usual limit of* f *at the point* $+\infty$;

ii) *for any $f \in \Lambda_{+\infty}$, $\partial f(+\infty) = f(+\infty)$.*

Proof: The uniqueness is clear. With regard to existence, if $f = \partial^n F = \partial^m G$ with $F, G \in \mathcal{C}(\mathbf{I}; +\infty)$ and $m, n \in \mathbb{N}$, $m \le n$, we will have:

$$\partial^n F = \partial^n \tilde{\mu}^{n-m} G$$

and therefore there will exist constants $c_1, c_2, \ldots, c_m$ such that

$$F(x) - \tilde{\mu}^{n-m} G(x) = \frac{1}{x} \left[c_1 + c_2 \log x + \ldots + c_n \log^{n-1} x \right].$$

Making $x \to +\infty$, we will obtain:

$$F(+\infty) - \left(\tilde{\mu}^{n-m} G \right)(+\infty) = 0$$

or

$$F(+\infty) = G(+\infty),$$

which proves that under the stated conditions we can put by definition $f(+\infty) = F(+\infty)$.
$\square$

The number $f(+\infty)$ associated in this way with the distribution $f \in \Lambda_{+\infty}(\mathbf{I})$ is naturally called *the limit of the distribution f at $+\infty$.*

For any function Φ, with indefinite integral over $\mathbf{I}$ and such that $\lim_{x \to +\infty} \Phi(x) = \gamma$ (in the usual sense), it is obvious that we also have $\Phi(+\infty) = \gamma$ in the sense of distributions.

Consider another example: if $\mathbf{I} =]0, +\infty[$ and α is a positive real number, from the equalities

$$\partial \left(\frac{1}{x^\alpha} \cos x^\alpha \right) = \mathbf{D} \left(\frac{1}{x^{\alpha-1}} \cos x^\alpha \right) = -\frac{\alpha - 1}{x^\alpha} \cos x^\alpha - \alpha \sin x^\alpha$$

it readily follows that, for any $x > 0$:

$$\sin x^\alpha = -\frac{1}{\alpha} \left[(\alpha - 1) \frac{\cos x^\alpha}{x^\alpha} + \partial \left(\frac{\cos x^\alpha}{x^\alpha} \right) \right].$$

Now, since the function $x^{-\alpha} \cos x^\alpha$ is an element of $\mathcal{C}(\mathbf{I}; +\infty) \subset \Lambda_{+\infty}(\mathbf{I})$ which has limit zero at infinity, it follows that $\sin x^\alpha \in \Lambda_{+\infty}(\mathbf{I})$, and that (in the sense of distributions):

$$\lim_{x \to +\infty} \sin x^\alpha = 0, \qquad (\alpha > 0).$$

Similarly it can be verified that, for any $\alpha > 0$, we have:

$$\lim_{x \to +\infty} \cos x^\alpha = 0.$$

Coming again to the case $\mathbf{I} =]a, +\infty[$, with $a \ge 0$, it is easy to see that, if f, g are in $\Lambda_{+\infty}(\mathbf{I})$ and α, β are two complex numbers then $\alpha f + \beta g$ also belongs to $\Lambda_{+\infty}(\mathbf{I})$. Moreover it is also easy to establish the equation

$$\lim_{x \to +\infty} (\alpha f + \beta g)(x) = \alpha \lim_{x \to +\infty} f(x) + \beta \lim_{x \to +\infty} g(x).$$

Before studying the relations between the limit and the operations of multiplication and change of variable, it is convenient to introduce the concept of *multiplier*. We have already seen (in section 1.6) that, under certain conditions, the product of a function φ and a distribution f will be a well defined distribution. If $f \in \mathcal{C}_\infty(\mathbf{I})$ is convergent at $+\infty$ then a function φ will be called a multiplier (at $+\infty$) if it satisfies conditions that ensure that φf is also a distribution convergent at $+\infty$. More precisely we shall say that a function φ, *of class $\mathcal{C}^\infty$ on $\mathbf{I}$ and with finite limit when $x \to +\infty$, is a multiplier at the point $+\infty$* (or, for the moment, simply a *multiplier*) if and only if, for every integer $p \geq 1$, we have, *in the usual sense*:

$$\lim_{x \to +\infty} x^p \varphi^{(p)}(x) \;=\; 0.$$

The simplest examples of multipliers are the constant functions on $\mathbf{I}$ and the functions of the form x^α, with $\alpha < 0$; also, any rational function $\varphi(x) = P(x)/Q(x)$ where $P(x)$ and $Q(x)$ are polynomials and the degree of $P(x)$ is not greater than the degree of $Q(x)$, is a multiplier on any interval of the form $]a, +\infty[$ $(a \geq 0)$ which does not contain any zero of $Q(x)$.

$\mathcal{M}_{+\infty}(\mathbf{I})$ will denote the set of all multipliers at the point $+\infty$ which are defined on $\mathbf{I}$; when the interval is fixed we may simply write $\mathcal{M}_{+\infty}$ instead of $\mathcal{M}_{+\infty}(\mathbf{I})$ provided no confusion is likely to arise.

Without any difficulty it may be verified that the sum of two multipliers and the product of a multiplier by a scalar are still multipliers; moreover, if φ and ψ are multipliers, the fact that $\varphi\psi \in \mathcal{C}^\infty(\mathbf{I}) \cap \mathcal{C}(\mathbf{I}; +\infty)$ and the equality

$$x^p(\varphi\psi)^{(p)}(x) \;=\; \sum_{k=0}^{p} \binom{p}{k} \left[x^k \varphi^{(k)}(x)\right] \left[x^{p-k} \psi^{(p-k)}(x)\right],$$

show that the product $\varphi\psi$ is again a multiplier; this says that $\mathcal{M}_{+\infty}(\mathbf{I})$ is a *subalgebra* of the *algebra* $\mathcal{C}^\infty \cap \mathcal{C}(\mathbf{I}; +\infty)$.

It is also easy to verify that, if $\varphi \in \mathcal{M}_{+\infty}$ then also $\partial\varphi \in \mathcal{M}_{+\infty}$. It is enough to note that, under the hypothesis, $\partial\varphi$ is certainly an infinitely differentiable function with finite limit when $x \to +\infty$ in the usual sense (equal to the limit of φ since $\hat{x}\varphi'(\hat{x})$ is infinitesimal) and to take into account that

$$x^p(\partial\varphi)^{(p)}(x) \;=\; x^p \mathbf{D}^{p+1}(x\varphi(x)) \;=\; x^{p+1}\varphi^{(p+1)}(x) + (p+1)x^p\varphi^{(p)}(x).$$

We may now state the following proposition:

Proposition 2.15 *If φ is a multiplier and f is a distribution which is convergent at $+\infty$ then φf is convergent at $+\infty$ and $(\varphi f)(+\infty) = \varphi(+\infty)f(+\infty)$.*

Proof: The proposition clearly holds in the particular case when $f \in \mathcal{C}(\mathbf{I}; +\infty)$. Therefore to complete the proof it will be enough to show that whenever a distribution f is such that with *any* function $\theta \in \mathcal{M}_{+\infty}(\mathbf{I})$

i) $\theta f \in \Lambda_{+\infty}(\mathbf{I})$, and

ii) $(\theta f)(+\infty) = \theta(+\infty)f(+\infty)$

the same will hold for the distribution ∂f.

If we set $g = \partial f$, and suppose that f does in fact satisfy the above conditions, we will have:

$$\varphi g \;=\; \varphi \partial f \;=\; \partial(\varphi f) - \psi f,$$

where $\psi = \partial \varphi - \varphi$ is an element of $\mathcal{M}_{+\infty}(\mathbf{I})$ (provided $\varphi \in \mathcal{M}_{+\infty}(\mathbf{I})$ as we had supposed). Then since the right-hand side is an element of $\Lambda_{+\infty}(\mathbf{I})$, the same is true for φg. Hence, taking into account the fact that $\psi(+\infty) = \partial\varphi(+\infty) - \varphi(+\infty) = 0$ we will have

$$\begin{aligned}
(\varphi g)(+\infty) \;&=\; \partial(\varphi f)(+\infty) - (\psi f)(+\infty) \\
&=\; \varphi(+\infty)f(+\infty) - \psi(+\infty)f(+\infty) \;=\; \varphi(+\infty)g(+\infty)
\end{aligned}$$

as was to be proved. $\square$

Before continuing, it is convenient to note here a small extension of the result stated in proposition 2.15 which the reader may easily verify: in the case when the distribution f is representable in the form $f = \partial^k F$, with $F \in \mathcal{C}(\mathbf{I}; +\infty)$ and $k \in \mathbb{N}$, to ensure that $\varphi f \in \Lambda_{+\infty}(\mathbf{I})$ and that $(\varphi f)(+\infty) = \varphi(+\infty)f(+\infty)$ it is sufficient that the function φ belongs to $\mathcal{C}^k(\mathbf{I}) \cap \mathcal{C}(\mathbf{I}; +\infty)$ and, for each positive integer number $p \le k$, satisfies the condition:

$$\lim_{x \to +\infty} x^p \varphi^{(p)}(x) \;=\; 0.$$

As an application we will prove now that, for any $\gamma \in \mathbb{R}$, we have:

$$\lim_{x \to +\infty} x^\gamma \sin x = 0 \quad \text{and} \quad \lim_{x \to +\infty} x^\gamma \cos x = 0.$$

Since, as seen above, $\sin x$ and $\cos x$ converge to 0 when $x \to +\infty$ (in the sense of the distributions), the result is an immediate consequence of the proposition 2.15 when x^γ is a multiplier, that is for $\gamma \le 0$.

Now let $\gamma > 0$. To begin with consider the case when γ is a positive integer, so that we can appeal to mathematical induction. Taking into account that the result is true for $\gamma = 0$, suppose it is true for $\gamma = p$ and consider the relations:

$$\begin{aligned}
\partial(x^p \sin x) \;&=\; \mathbf{D}(x^{p+1} \sin x) \;=\; (p+1)x^p \sin x + x^{p+1} \cos x, \\
\partial(x^p \cos x) \;&=\; \mathbf{D}(x^{p+1} \cos x) \;=\; (p+1)x^p \cos x - x^{p+1} \sin x.
\end{aligned}$$

Since by the induction hypothesis, $x^p \sin x$ and $x^p \cos x$, and hence $\partial(x^p \sin x)$ and $\partial(x^p \cos x)$, all converge to 0 when $x \to +\infty$, it follows that the same is true for $x^{p+1} \cos x$ and $x^{p+1} \sin x$. Thus the result that we want to prove is true when γ is an integer. For positive non-integral values of γ, it is enough to take into account again the proposition 2.15 and to consider the equalities:

$$\begin{aligned}
x^\gamma \sin x \;&=\; x^{\gamma - c(\gamma) - 1}\big[x^{1+c(\gamma)} \sin x\big], \\
x^\gamma \cos x \;&=\; x^{\gamma - c(\gamma) - 1}\big[x^{1+c(\gamma)} \cos x\big],
\end{aligned}$$

where $c(\gamma)$ denotes the greatest integer not greater than γ.

Proposition 2.16 *Consider the intervals* $\mathbf{I} =]a,+\infty[$ *and* $\mathbf{J} =]c,+\infty[$ *(where $0 \le a, c < +\infty$) and let $\varphi : \mathbf{J} \to \mathbf{I}$ be a C^∞-function such that:*

α) $\lim_{t\to+\infty} \varphi(t) = +\infty$,

β) $\varphi'(t) \ne 0$ *for each* $t \in \mathbf{J}$,

γ) $\frac{\varphi(t)}{t\varphi'(t)} \in \mathcal{M}_{+\infty}(\mathbf{J})$.

If f is a distribution defined on $\mathbf{I}$ and convergent at $+\infty$, then $f \circ \varphi$ is a distribution (defined on $\mathbf{J}$) which is also convergent at $+\infty$ and moreover, we have $(f \circ \varphi)(+\infty) = f(+\infty)$.

Proof: Since, in the case when f is an element of $\mathcal{C}(\mathbf{I}; +\infty)$ we have, under the hypothesis, that i) $f \circ \varphi \in \Lambda_{+\infty}(\mathbf{J})$, and ii) $(f \circ \varphi)(+\infty) = f(+\infty)$, it remains to prove that if a distribution $f \in \Lambda_{+\infty}(\mathbf{I})$ satisfies i) and ii), the same holds for $g = \partial f$.

Now (denoting by ∂^* the mapping from $\mathcal{C}_\infty(\mathbf{J})$ into itself such that $\partial^* h(t) = \mathbf{D}_t[th(t)]$ for each $h \in \mathcal{C}_\infty(\mathbf{J})$) we have:

$$
\begin{aligned}
(g \circ \varphi)(t) &= \frac{1}{\varphi'(t)} \mathbf{D}_t[\varphi(t)(f \circ \varphi)(t)] \\
&= (f \circ \varphi)(t) + \frac{\varphi(t)}{t\varphi'(t)} [\partial^*(f \circ \varphi)(t) - (f \circ \varphi)(t)].
\end{aligned}
$$

By the hypothesis, $\varphi(t)/t\varphi'(t)$ is a multiplier and so the last term will converge to 0 when $t \to +\infty$, while $(f \circ \varphi)(t)$ will converge to $(f \circ \varphi)(+\infty) = f(+\infty) = g(+\infty)$ due to the hypothesis made on f. It follows then that $g \circ \varphi$ converges at the point $+\infty$ and that we have $(g \circ \varphi)(+\infty) = g(+\infty)$ and this ends the proof. $\square$

The proposition 2.16 may, of course, also be slightly extended in a way similar to the one referred to in connection with the proposition 2.15: if $f = \partial^k F$ ($F \in \mathcal{C}(\mathbf{I}; +\infty)$, $k \in \mathbb{N}$), to obtain the same result it is enough to suppose that the function $\varphi : \mathbf{J} \to \mathbf{I}$ satisfies the conditions:

$$
\lim_{t\to+\infty} \varphi(t) = +\infty, \qquad \frac{1}{\varphi'} \in \mathcal{C}^k(\mathbf{J})
$$

and that, by setting $\psi(t) = \varphi(t)/t\varphi'(t)$, we have

$$
\lim_{t\to+\infty} t^p \psi^{(p)}(t) = 0,
$$

for $p = 1, 2, \ldots, k$.

As examples of functions φ satisfying the conditions referred to in the statement of proposition 2.16, we may quote the powers t^α (with $\alpha > 0$) and the exponentials a^t (with $a > 1$). Thus, taking into account that, as we have already seen

$$
\lim_{x\to+\infty} x^\gamma \sin x = 0 \qquad (\gamma \in \mathbb{R})
$$

it follows for example that, for any real γ:

$$
\lim_{x\to+\infty} e^{\gamma x} \sin e^x = 0.
$$

The function $\varphi(t) = \log t$ does not satisfy the third condition of the statement of proposition 2.16; we remark that the distribution $f(x) = \sin \log x$ does not converge when $x \to +\infty$. In fact, observe that, if this distribution were in $\Lambda_{+\infty}(\mathbf{I})$, (with $\mathbf{I} =]0, +\infty[$, for example), then there would exist a function $F \in \mathcal{C}(\mathbf{I}; +\infty)$ and a number $p \in \mathbb{N}$ such that $f = \partial^p F$. It would then be easy to see that, under this hypothesis, there should also exist, for every $n > p$ a function $F_n \in \mathcal{C}(\mathbf{I}; +\infty)$ such that $f = \partial^n F_n$. But by a simple calculation it can be seen that we have for any $m \in \mathbb{N}$:

$$\partial^{(4m)}(f) = (-1)^m 4^m f,$$

from which we may deduce that all functions g_m satisfying the condition $\partial^{(4m)} g_m = f$ may be represented in the form:

$$g_m(x) = (-1)^m \frac{1}{4^m} \sin \log x + \pi_m(x),$$

where π_m is a function[4] with limit zero when $x \to +\infty$. Since for all $m \in \mathbb{N}$ none of the functions g_m belongs to the space $\mathcal{C}(\mathbf{I}; +\infty)$, it follows that $f(x) = \sin \log x$ does not converge in the sense of distributions when $x \to +\infty$.

The notions of convergence and of limit at the point $+\infty$ have so far been introduced for distributions defined on an interval contained in $]0, +\infty[$. However in the case when f is a distribution defined in an arbitrary interval, $\mathbf{I}$, unbounded above, it is natural to say that f is convergent at $+\infty$, and to write $f \in \Lambda_{+\infty}(\mathbf{I})$, if and only if the restriction of f to the interval $\mathbf{J} = \mathbf{I} \cap]0, +\infty[$ is an element of $\Lambda_{+\infty}(\mathbf{J})$. In such circumstances we shall write, by definition, $f(+\infty) = f_{|\mathbf{J}}(+\infty)$.

It should not be necessary to state the corresponding definitions of convergence and limit at the point $-\infty$ for a distribution defined on an interval unbounded below. These definitions are easily reduced to those which we have been studying, since the distribution $f(x)$, defined on $] -\infty, b[$, converges when $x \to -\infty$ if and only if the distribution $g(x) = f(-x)$ (defined on $] - b, +\infty[$) converges when $x \to +\infty$, and we have in this case $f(-\infty) = g(+\infty)$.

Finally we will say that $f \in \mathcal{C}_\infty(\mathbb{R})$ converges at ∞ if and only if it converges at $+\infty$ and $-\infty$ and, moreover, we have $f(+\infty) = f(-\infty)$; this common value may then be denoted simply by $f(\infty)$. For example, if m and n are positive integers and $f(x) = x^m \sin x^n$ we have $f(\infty) = 0$.

The last part of this section will be devoted to the study of the concept of the limit of a distribution at a point $a \in \mathbb{R}$, which is a limit point of its domain. It should

---

[4] The function π_m will be necessarily of the form

$$\pi_m(x) = \frac{1}{x} \sum_{p=0}^{4m-1} c_p \log^p x,$$

where $c_0, \ldots, c_{4m-1}$ are constants.

be enough simply to adapt the preceeding definitions and results to the new situation leaving the proofs to the reader. Hence what follows will only be a brief adaptation of the theory developed in section 2.1 and during the first part of this section.

If $\mathbf{I} =]a, b[$ (with $-\infty < a < b \leq +\infty$) we will denote by $\mathcal{C}(\mathbf{I}; a)$ the subset of $\mathcal{C}(\mathbf{I})$ comprising all functions which have (finite) limit when $x \to a$. Setting, for each $F \in \mathcal{C}(\mathbf{I}; a)$ and each $x \in \mathbf{I}$:

$$\mu_a F(x) \;=\; \frac{1}{x-a} \int_a^x F(t)dt,$$

we will certainly have $\mu_a F \in \mathcal{C}(\mathbf{I}; a)$ and

$$\lim_{x \to a} \mu_a F(x) \;=\; \lim_{x \to a} F(x).$$

Defining the operator $\partial_a : \mathcal{C}(\mathbf{I}) \to \mathcal{C}(\mathbf{I})$ as in section 2.1, we will now say that a distribution $f \in \mathcal{C}_\infty(\mathbf{I})$ is *convergent at the point* a, or that $f(x)$ has a limit when $x \to a$, if and only if there exist $F \in \mathcal{C}(\mathbf{I}; a)$ and $n \in \mathbb{N}$ such that $f = \partial_a^n F$.

We will denote by $\Lambda_a(\mathbf{I})$ the subset of $\mathcal{C}_\infty(\mathbf{I})$ comprising all distributions which are convergent at the point a. For example, if $\mathbf{J}$ is an interval such that the point a is an interior point of $\mathbf{J}$ and f is a distribution defined on $\mathbf{J}$ which is continuous at the point a, the restriction of f to $\mathbf{J} \cap]a, +\infty[$ will be a distribution convergent when $x \to a$.

It is easily seen that $\Lambda_a(\mathbf{I})$ is a linear subspace of $\mathcal{C}(\mathbf{I})$, the restriction of ∂_a to $\Lambda_a(\mathbf{I})$ being an automorphism of this space; and also that there exists a unique mapping $f \rightsquigarrow \lim_{x \to a} f(x) = \lim_a f$ from $\Lambda_a(\mathbf{I})$ into $\mathbb{C}$ satisfying the following conditions:

i) if $f \in \mathcal{C}(\mathbf{I}; a)$, $\lim_a f$ is the usual limit of $f(x)$ when $x \to a$;

ii) for any $f \in \Lambda_a(\mathbf{I})$, $\lim_a (\partial_a f) = \lim_a f$.

The definition of limit at the point a for an arbitrary distribution $f \in \Lambda_a(\mathbf{I})$ will be based on this last proposition. It is easily proved that if $f = \partial_a^n F$, with $F \in \mathcal{C}(\mathbf{I}; a)$, and $n \in \mathbb{N}$, then we can set

$$\lim_{x \to a} f(x) \;=\; \lim_{x \to a} F(x).$$

The linearity of the mapping $f \rightsquigarrow \lim_a f$ is also immediate.

Further, if we agree to call those functions $\varphi \in \mathcal{C}^\infty(\mathbf{I}) \cap \mathcal{C}(\mathbf{I}; a)$ which, for every positive integer p, satisfy the condition

$$\lim_{x \to a} (x-a)^p \varphi^{(p)}(x) \;=\; 0$$

multipliers at the point a (defined on the interval $\mathbf{I}$), and to denote by $\mathcal{M}_a(\mathbf{I})$ the set of all such functions, we can state the following propositions:

Proposition 2.17 *If φ is a multiplier at the point a and f is a distribution which converges at the same point, φf is again convergent at the point a and, moreover we have:*

$$\lim_{x \to a} (\varphi f)(x) \;=\; \lim_{x \to a} \varphi(x) \lim_{x \to a} f(x).$$

Proposition 2.18 *Let* $\mathbf{I} =]a, b[$ *and* $\mathbf{J} =]c, d[$ *be two intervals of* $\mathbb{R}$ *and* $\varphi : \mathbf{J} \to \mathbf{I}$ *a function of class* C^∞ *satisfying the following conditions:*

$\alpha)$ $\lim_{t \to c} \varphi(t) = a$;

$\beta)$ $\varphi'(t) \neq 0$, *for every* $t \in \mathbf{J}$;

$\gamma)$ $\frac{\varphi(t) - a}{(t-c)\varphi'(t)} \in \mathcal{M}_c(\mathbf{J})$.

If $f \in \Lambda_a(\mathbf{I})$, *then* $f \circ \varphi \in \Lambda_c(\mathbf{J})$ *and moreover we have:*

$$\lim_{t \to c}(f \circ \varphi)(t) \;=\; \lim_{x \to a} f(x).$$

Each of these two propositions may be given extensions similar to those mentioned with respect to propositions 2.15 amd 2.16.

As examples of multipliers at the point a (defined on the interval $]a, b[$) it is convenient to mention the restrictions to $\mathbf{I}$ of the functions of class C^∞, defined on an open set containing the interval $[a, b[$.

Any change of variables which transforms bounded intervals into unbounded intervals, and conversely, may be treated in an entirely similar way; thus, for example, it is possible to prove, in a similar fashion to the proof of 2.16, the following proposition:

Proposition 2.19 *Let* $\mathbf{I} =]a, +\infty[$, $\mathbf{J} =]c, d[$ *and* $\varphi : \mathbf{J} \to \mathbf{I}$ *a function of class* C^∞ *satisfying the following conditions:*

$\alpha)$ $\lim_{t \to c} \varphi(t) = +\infty$;

$\beta)$ $\varphi'(t) \neq 0$, *for each* $t \in \mathbf{J}$;

$\gamma)$ $\frac{\varphi(t)}{(t-c)\varphi'(t)} \in \mathcal{M}_c(\mathbf{J})$.

If $f \in \Lambda_{+\infty}(\mathbf{I})$, *then* $f \circ \varphi \in \Lambda_c(\mathbf{J})$ *and, moreover*

$$\lim_{t \to c}(f \circ \varphi)(t) \;=\; f(+\infty).$$

An analogous proposition also holds concerning a change of variable $t = \psi(x)$ which transforms the interval $\mathbf{I}$ into the interval $\mathbf{J}$ and is such that $\psi(x) \to c$ as $x \to +\infty$, $\psi'(x) \neq 0$ for each $x \in \mathbf{I}$ and

$$\frac{\psi(x) - c}{x\psi'(x)} \in \mathcal{M}_{+\infty}(\mathbf{I}).$$

In such a case it may be concluded that if $g \in \Lambda_c(\mathbf{J})$ then $g \circ \psi \in \Lambda_{+\infty}(\mathbf{I})$, and $(g \circ \psi)(+\infty) = \lim_{t \to c} g(t)$.

Another result which may be obtained by a method identical to that used in the proof of the proposition 2.12 is the following:

Proposition 2.20 *Let* $\mathbf{I} =\,]a,b[$, $\mathbf{I}' =\,]a,b'[$ *and* $b' < b$; *if* $f \in \mathcal{C}_\infty(\mathbf{I})$ *and* $f_{|\mathbf{I}'} \in \Lambda_a(\mathbf{I}')$ *then* $f \in \Lambda_a(\mathbf{I})$ *and*

$$\lim_{x \to a} f(x) = \lim_{x \to a} f_{|\mathbf{I}'}(x).$$

From this there follows at once:

Proposition 2.21 *If* $\mathbf{I}$ *and* $\mathbf{I}'$ *are as in the previous proposition,* f *and* g *are two distributions on* $\mathbf{I}$ *and* $f = g$ *on* $\mathbf{I}'$ *then* $f \in \Lambda_a(\mathbf{I})$ *if and only if* $g \in \Lambda_a(\mathbf{I})$ *and in this case*

$$\lim_{x \to a} f(x) \;=\; \lim_{x \to a} g(x).$$

Finally, by making a change of variable of the same type as that considered in the proposition 2.19, the "local character" of the limits at infinity is easily recognised; for example, we have the following:

Proposition 2.22 *Let* $a, a' \in \mathbb{R}$ *with* $a < a'$ *and let* $\mathbf{I}$ *and* $\mathbf{I}'$ *denote the intervals* $]a, +\infty[$ *and* $]a', +\infty[$, *respectively. For a distribution* f, *defined on* $\mathbf{I}$, *to be convergent at the point* $+\infty$ *it is necessary and sufficient that its restriction to* $\mathbf{I}'$ *be itself convergent; if* g *is another distribution defined on* $\mathbf{I}$ *and if* $f = g$ *on* $\mathbf{I}'$, g *will be convergent at* $+\infty$ *if and only if* f *is convergent and in this case we will have* $f(+\infty) = g(+\infty)$.

It goes without saying that the concepts of convergence and limit at a point b, for a distribution defined on $\mathbf{I} =\,]a, b[$ $(-\infty \le a < b < +\infty)$ may be defined in an entirely similar manner; and it is obvious that such concepts have similar properties to those described in the propositions 2.17 to 2.22.

Suppose now that f is a distribution defined on an interval $\mathbf{I}$ and that $a \in \mathbb{R}$ is a limit point of $\mathbf{I}$, other than the upper limit of that interval. We will then say that f has a *limit on the right at the point* a if and only if the restriction of f to $\mathbf{I}_2 = \mathbf{I} \cap\,]a, +\infty[$ is convergent at the point a; in this case, the limit of $f_{|\mathbf{I}_2}$ at the point a will be, by definition, the limit on the right of f at the same point and we may therefore write

$$f(a^+) \;=\; \lim_{x \to a+} f(x) \;=\; \lim_{x \to a} f_{|\mathbf{I}_2}.$$

Similarly there may be defined the limit on the left at a point a (with respect to a distribution defined on an interval of which the point a is not the lower bound and is a limit point).

Finally, if $f \in \mathcal{C}_\infty(\mathbf{I})$ and a is an interior point of the interval $\mathbf{I}$, we will say that f is *convergent at the point* a or that f *has a limit at the point* a if and only if the two one-sided limits, $f(a^+)$ and $f(a^-)$ both exist and are equal; in this case the limit of $f(x)$ when $x \to a$ will be, by definition, the common value:

$$\lim_{x \to a} f(x) \; = \; f(a^+) \; = \; f(a^-).$$

In this way we have defined the concepts of convergence and limit of a distribution at any point which is a limit point of the interval on which the distribution is defined.

For example, if a is an arbitrary point of $\mathbb{R}$, then any linear combination of derivatives of $\delta_{(a)}$, (which will be the null distribution both on $]-\infty, a[$ and on $]a, +\infty[$), will have limit zero at the point a (as well as at any other point); we already know that among all such distributions there is only one which is continuous at the point a, namely the null distribution.

The reader should have no difficulty in extending the properties of the concept of limit established previously to the cases of lateral limits and of limits at interior points of the interval on which a distribution is defined.

2.3   Extensions Revisited. Trivial Extension. Truncation.

From the previous section it follows immediately that a distribution continuous at a point a has a limit at the same point, and moreover the equality

$$\lim_{x \to a} f(x) \; = \; f(a)$$

holds. Hence, if a is a point interior to the interval $\mathbf{I}$, $\mathbf{I}_1 = \mathbf{I} \cap]-\infty, a[$, $\mathbf{I}_2 = \mathbf{I} \cap]a, +\infty[$, $f_1 \in \mathcal{C}_\infty(\mathbf{I}_1)$ and $f_2 \in \mathcal{C}_\infty(\mathbf{I}_2)$, it is obvious that, for f_1 and f_2 to be extendable into a distribution f which is continuous at the point a, it is necessary that both distributions be convergent when $x \to a$ and also that

$$\lim_{x \to a} f_1(x) \; = \; \lim_{x \to a} f_2(x).$$

Conversely, if f_1 and f_2 satisfy this condition, it will be possible to represent them in the form $f_1 = \partial_a^n F_1$, $f_2 = \partial_a^n F_2$ with $F_1 \in \mathcal{C}(\mathbf{I}_1; a)$, $F_2 \in \mathcal{C}(\mathbf{I}_2; a)$ and

$$\lim_{x \to a} F_1(x) \; = \; \lim_{x \to a} f_1(x) \; = \; \lim_{x \to a} f_2(x) \; = \; \lim_{x \to a} F_2(x).$$

If we denote by F the continuous function on $\mathbf{I}$ which is such that $F_{|_{\mathbf{I}_1}} = F_1$ and $F_{|_{\mathbf{I}_2}} = F_2$, then the distribution $f = \partial_a^n F$ will be the common extension of f_1 and f_2 to $\mathbf{I}$, which is continuous at the point a; and, as we already know, this (continuous) extension will be unique. We can therefore state:

Proposition 2.23 *Let a be a point belonging to the interior of the interval* $\mathbf{I}$, *and* $\mathbf{I}_1 = \mathbf{I} \cap]-\infty, a[$, $\mathbf{I}_2 = \mathbf{I} \cap]a, +\infty[$, $f_1 \in \mathcal{C}_\infty(\mathbf{I}_1)$ *and* $f_2 \in \mathcal{C}_\infty(\mathbf{I}_2)$. *For the existence of a common extension of f_1 and f_2 to* $\mathbf{I}$ *which is continuous at the point a to exist, it is necessary and sufficient that f_1 and f_2 both converge to the same limit when $x \to a$; under such conditions the extension is unique.*

In the sequel, the extension of f_1 and f_2 under the conditions of Proposition 2.23 will be called the *continuous extension of the pair* (f_1, f_2); when a pair (f_1, f_2) admits a continuous extension it will also be said to be a *continuously extendable pair*. The terms "continuous" and "continuously" may be omitted when there is no danger of confusion. The following results are immediate consequences of the propositions stated in sections 2.1 and 2.2:

Proposition 2.24 *If α, β are complex numbers and (f_1, f_2) and (g_1, g_2) are two extendable pairs (with $f_1, g_1 \in C_\infty(\mathbf{I}_1)$ and $f_2, g_2 \in C_\infty(\mathbf{I}_2)$) then the pair defined by $(\alpha f_1 + \beta g_1, \alpha f_2 + \beta g_2)$ is also extendable and its extension is $\alpha f + \beta g$ where f and g are the extensions of (f_1, f_2) and (g_1, g_2), respectively.*

Proposition 2.25 *Let $\varphi \in C^\infty(\mathbf{I})$, $\varphi_1 = \varphi|_{\mathbf{I}_1}$, $\varphi_2 = \varphi|_{\mathbf{I}_2}$, $f_1 \in C_\infty(\mathbf{I}_1)$ and $f_2 \in C_\infty(\mathbf{I}_2)$; if the pair (f_1, f_2) is extendable, with extension f, the same will be true for the pair $(\varphi_1 f_1, \varphi_2 f_2)$ and its extension will be φf.*

Proposition 2.26 *Let $\mathbf{I}$ and $\mathbf{J}$ be two intervals of $\mathbb{R}$, $\varphi : \mathbf{J} \to \mathbf{I}$ a C^∞-function such that $\varphi'(t) > 0$ for each $t \in \mathbf{J}$,[5] and c an interior point of $\mathbf{J}$. Further let $a = \varphi(c)$, $\mathbf{I}_1 = \mathbf{I} \cap] -\infty, a[$, $\mathbf{I}_2 = \mathbf{I} \cap]a, +\infty[$, $\mathbf{J}_1 = \mathbf{J} \cap] -\infty, c[$, $\mathbf{J}_2 = \mathbf{J} \cap]c, +\infty[$, $\varphi_1 = \varphi|_{\mathbf{J}_1}$, $\varphi_2 = \varphi|_{\mathbf{J}_2}$, $f_1 \in C_\infty(\mathbf{I}_1)$, $f_2 \in C_\infty(\mathbf{I}_2)$ and, finally, $g_1 = f_1 \circ \varphi_1$, $g_2 = f_2 \circ \varphi_2$. If the pair (f_1, f_2) is extendable, with extension f, the same holds also for the pair (g_1, g_2). If g denotes the corresponding extension, then $g = f \circ \varphi$.*

As already seen, the uniqueness of the continuous extension is a consequence of the fact that no linear combination of derivatives of $\delta_{(a)}$ (with the exception of the null combination) is continuous at the point a. It is, however, easy to see not only that such linear combinations cannot be continuous but also that their primitives (of first order) cannot be continuous. In fact, the primitives of the distribution

$$\sum_{k=0}^{m} c_k \delta^{(k)}(x - a),$$

(with $m \in \mathbb{N}$; $c_0, \ldots, c_m \in \mathbf{C}$ and $c_m \neq 0$) are precisely the distributions of the form

$$c + c_0 \mathbf{H}(x - a) + \sum_{k=1}^{m} c_k \delta^{(k-1)}(x - a)$$

with $c \in \mathbf{C}$. If $c_0 = 0$ this distribution is the sum of a constant with a non null linear combination of derivatives of $\delta_{(a)}$; if $c_0 \neq 0$, the primitives considered have different limits on the left and on the right of the point a. In neither case can they be continuous at that point.

---

[5] If $\varphi'(t) < 0$ for each $t \in \mathbf{J}$ the modifications to be introduced in this statement are obvious.

This fact suggests a statement of the extension problem for distributions, with unique solution, under more general conditions than extension by continuity. For this, if a is an interior point of the interval $\mathbf{I}$, let us agree to say that a distribution $f \in \mathcal{C}_\infty(\mathbf{I})$ is *precontinuous at the point* a if and only if there exists a distribution g continuous at a and such that $f = \mathbf{D}g$; denote by $W_a(\mathbf{I})$ the subspace of $\mathcal{C}_\infty(\mathbf{I})$ comprising all distributions which are precontinuous at a. Then $W_a(\mathbf{I})$ is the image of the space $V_a(\mathbf{I})$ by the linear map $\mathbf{D}$ and is, therefore, a linear space.

Proposition 2.9 shows that we have $V_a \subset W_a$. Clearly the inclusion is strict (for example, the restriction to $\mathbf{I}$ of the Heaviside distribution at the point a, $\mathbf{H}(x - a)$, is an element of $W_a \backslash V_a$).

If again we let $\mathbf{I}_1 = \mathbf{I} \cap\,]-\infty, a[$ and $\mathbf{I}_2 = \mathbf{I} \cap\,]a, +\infty[$, we will say that a distribution $f_1 \in \mathcal{C}_\infty(\mathbf{I}_1)$ is *preconvergent at the point* a if and only if there exists a distribution g_1, defined on $\mathbf{I}_1$, which converges when $x \to a$, and is such that $f_1 = \mathbf{D}g_1$; and similarly for a distribution $f_2 \in \mathcal{C}(\mathbf{I}_2)$. We have therefore

> **Proposition 2.27** *Under the same conditions as those of Proposition 2.23, an extension of the pair (f_1, f_2) to $\mathbf{I}$ which is precontinuous at the point a (that is, a distribution f defined on $\mathbf{I}$, precontinuous at a and such that $f_{|\mathbf{I}_1} = f_1$, $f_{|\mathbf{I}_2} = f_2$) exists if and only if f_1 and f_2 are preconvergent at a. When this condition is satisfied the extension is unique.*

Proof: It is clear that a precontinuous extension cannot exist if f_1 or f_2 are not preconvergent; if both are preconvergent then choosing two distributions $g_1 \in \Lambda_a(\mathbf{I}_1)$, $g_2 \in \Lambda_a(\mathbf{I}_2)$ such that $f_1 = \mathbf{D}g_1$, $f_2 = \mathbf{D}g_2$ and

$$\lim_{x \to a} g_1(x) \;=\; \lim_{x \to a} g_2(x)$$

(which is always possible) and denoting by g the continuous extension of the pair (g_1, g_2), we will have that $f = \mathbf{D}g$ is a precontinuous extension for the pair (f_1, f_2). The uniqueness of the extension is a consequence of the fact that the null distribution is the unique distribution on $\mathbf{I}$ which is null on $\mathbf{I}_1$ and $\mathbf{I}_2$ and is precontinuous at a. $\qquad\square$

Clearly when f_1 and f_2 are convergent at a (and therefore also preconvergent at the same point) the precontinuous extension of the pair (f_1, f_2) coincides with its continuous extension. However it may happen that a precontinuous extension exists without a continuous extension: for example, the distribution $\mathbf{H}$ is the precontinuous extension to $\mathbb{R}$ of the pair $(\mathbf{0}, \mathbf{1})$, where $\mathbf{0}$ is the null distribution defined on $]-\infty, 0[$ and $\mathbf{1}$ is the unit distribution on $]0, +\infty[$; and it is obvious that for this pair there does not exist any extension which is continuous at the origin.

We now remark that the equality

$$\varphi \mathbf{D}g \;=\; \mathbf{D}(\varphi g) - \varphi' g$$

and other previous results (in particular that of Proposition 2.7), allow us to recognise that if $f = \mathbf{D}g$ is a distribution precontinuous at the point a then its product by a

$\mathcal{C}^\infty$-function is also precontinuous at the same point; similarly, from

$$\mathbf{D}g \circ \varphi \;=\; \frac{1}{\varphi'}\, \mathbf{D}(g \circ \varphi)$$

it follows easily that the change of variable transforms distributions which are precontinuous at a certain point into distributions which are precontinuous at the corresponding point.

Taking all these facts into account, it may be concluded without any difficulty that the propositions 2.24, 2.25 and 2.26 stated above for continuous extensions remains valid if the words "extension" and "extendable" are understood in the sense of the precontinuous extensions. To prove 2.25, for example, it would be enough to observe that, if φ is a function of the class $\mathcal{C}^\infty$ and if f is the precontinuous extension of the pair (f_1, f_2) then the distribution φf, precontinuous at a and equal to $\varphi_1 f_1$ on $\mathbf{I}_1$ and to $\varphi_2 f_2$ on $\mathbf{I}_2$, will certainly be the precontinuous extension of the pair $(\varphi_1 f_1, \varphi_2 f_2)$; and similarly for the other cases.

A particularly important case of a precontinuous extension is the so-called *trivial extension* of a distribution. Let $a \in \mathbb{R}$ and let f be a distribution on the interval $]a, +\infty[$; if f is preconvergent at the point a there will exist one (and only one) distribution $\check{f} \in \mathcal{C}_\infty(\mathbb{R})$ which is precontinuous at the point a and is such that

$$\check{f}_{|]a,+\infty[} = f \qquad \check{f}_{|]-\infty,a[} = \mathbf{0}.$$

In such a case we will say that the distribution f is *trivially extendable* and we will call the distribution $\check{f}$ the *trivial extension* of f (to $\mathbb{R}$). In a similar way we may define the trivial extension of a distribution f, defined on $]-\infty, b[$ and preconvergent at the point b. This extension will again be denoted by $\check{f}$. It is the unique distribution which extends f, is the null distribution on $]b, +\infty[$ and is precontinuous at the point b.

This notion of trivial extension just defined (which can be extended to more general situations to be considered later), will be useful on many ocasions. For the moment we will use it to define the *truncating* operation which has been used by Heaviside to solve several types of differential equations.

Given a point a fixed in $\mathbb{R}$, let f be a distribution defined on $\mathbb{R}$ and precontinuous at the point a. We will call the trivial extension (to $\mathbb{R}$) of the restriction of f to the interval $]a, +\infty[$ the *truncation of f to the left of a*, and will denote it by $f\mathbf{H}_{(a)}$. Similarly we will call the trivial extension of the restriction of f to the interval $]-\infty, a[$ the *truncation of f to the right of a*, and denote it by $f\mathbf{H}^*_{(a)}$.[6]

---

[6]The symbol $\mathbf{H}^*_{(a)}$ is used to denote the distribution $\mathbf{1} - \mathbf{H}_{(a)}$ where $\mathbf{H}_{(a)}$ is the Heaviside distribution at the point a. It is convenient to observe that the notations $f\mathbf{H}_{(a)}$ and $f\mathbf{H}^*_{(a)}$ now adopted suggest that the distributions considered (the truncations of f to the left and to the right of a) may in some sense be looked at as "products" of the distribution f by the distributions $\mathbf{H}_{(a)}$ and $\mathbf{H}^*_{(a)}$, although not in the sense of the general definition given earlier. We will refer again to such "products" in the sequel.

As $f\mathbf{H}_{(a)}$ and $f\mathbf{H}^*_{(a)}$ are precontinuous distributions at the point a their sum will also be a precontinuous distribution; and clearly this sum will coincide with f in each one of the intervals $]-\infty, a[$ and $]a, +\infty[$. Since by hypothesis f is also precontinuous at the point a, the uniqueness of the extension shows that we will necessarily have:

$$f\mathbf{H}_{(a)} + f\mathbf{H}^*_{(a)} \;=\; f.$$

Given the scalars $\alpha, \beta \in \mathbf{C}$ and the distributions f, g defined on $\mathbb{R}$ and precontinuous at the point a, the distribution $h = \alpha f + \beta g$ will also be precontinuous at the same point and will have a truncation $h\mathbf{H}_{(a)}$, which is the unique distribution equal to h on $]a, +\infty[$, null on $]-\infty, a[$ and precontinuous at a. Since the distribution $\alpha(f\mathbf{H}_{(a)}) + \beta(g\mathbf{H}_{(a)})$ enjoys the same properties it may be concluded that:

$$h\mathbf{H}_{(a)} \;=\; (\alpha f + \beta g)\mathbf{H}_{(a)} \;=\; \alpha\left(f\mathbf{H}_{(a)}\right) + \beta\left(g\mathbf{H}_{(a)}\right).$$

Similarly

$$(\alpha f + \beta g)\mathbf{H}^*_{(a)} \;=\; \alpha\left(f\mathbf{H}^*_{(a)}\right) + \beta\left(g\mathbf{H}^*_{(a)}\right).$$

In the same way if $\varphi \in \mathcal{C}^\infty(\mathbb{R})$ and $f \in W_a(\mathbb{R})$ then $\varphi f \in W_a(\mathbb{R})$ and

$$(\varphi f)\mathbf{H}_{(a)} = \varphi\left(f\mathbf{H}_{(a)}\right) \quad \text{and} \quad (\varphi f)\mathbf{H}^*_{(a)} = \varphi\left(f\mathbf{H}^*_{(a)}\right).$$

For a constant distribution, $g(x) = c$, the truncation $g\mathbf{H}_{(a)}$ is precisely the product of the constant c by the Heaviside distribution, $c\mathbf{H}_{(a)}$; in fact such a truncation is the unique distribution which is null on $]-\infty, a[$, equal to c on $]a, +\infty[$ and precontinuous at the point a, and clearly this is the case for the distribution $c\mathbf{H}_{(a)}$.

We may now state the following theorem:

Theorem 2.28 (Formula of Jumps) *If f is a distribution defined on $\mathbb{R}$ and continuous at the point a then*

$$\begin{aligned}
\mathbf{D}\left(f\mathbf{H}_{(a)}\right) &= (\mathbf{D}f)\,\mathbf{H}_{(a)} + f(a)\delta_{(a)} \\
\mathbf{D}\left(f\mathbf{H}^*_{(a)}\right) &= (\mathbf{D}f)\,\mathbf{H}^*_{(a)} - f(a)\delta_{(a)}.
\end{aligned} \tag{2.10}$$

More generally, if n is an integer greater than or equal to 1, $f \in \mathcal{C}_\infty(\mathbb{R})$ and $\mathbf{D}^{n-1}f$ is continuous at the point a then the following formulae

$$\begin{aligned}
\mathbf{D}^n\left(f\mathbf{H}_{(a)}\right) &= (\mathbf{D}^n f)\,\mathbf{H}_{(a)} + \sigma \\
\mathbf{D}^n\left(f\mathbf{H}^*_{(a)}\right) &= (\mathbf{D}^n f)\,\mathbf{H}^*_{(a)} - \sigma,
\end{aligned} \tag{2.11}$$

where $\sigma = \mathbf{D}^{n-1}f(a)\delta_{(a)} + \mathbf{D}^{n-2}f(a)\delta'_{(a)} + \ldots + f(a)\delta^{(n-1)}_{(a)}$, hold.

Proof: If f is continuous at the point a, the distribution $f - f(a)$ will have limit zero to the right of the point a, and therefore, in accordance with Proposition 2.23, there will exist a distribution continuous at the point a, null on $]-\infty, a[$ and equal to $f - f(a)$ on $]a, +\infty[$.

However by the uniqueness of the precontinuous extension, such a distribution must coincide with the truncation

$$[f - f(a)]\mathbf{H}_{(a)} = f\mathbf{H}_{(a)} - f(a)\mathbf{H}_{(a)}.$$

The derivative of this distribution, precontinuous at the point a, null on $]-\infty, a[$ and equal to $\mathbf{D}f$ on $]a, +\infty[$ must therefore be equal to the truncation of the derivative of f, $(\mathbf{D}f)\mathbf{H}_{(a)}$; that is, we must have

$$\mathbf{D}\left(f\mathbf{H}_{(a)} - f(a)\mathbf{H}_{(a)}\right) = (\mathbf{D}f)\mathbf{H}_{(a)}$$

or

$$\mathbf{D}\left(f\mathbf{H}_{(a)}\right) = (\mathbf{D}f)\mathbf{H}_{(a)} + f(a)\delta_{(a)}.$$

The second formula in (2.10) may be obtained similarly, or by noting that

$$\mathbf{D}\left(f\mathbf{H}_{(a)}^{*}\right) = \mathbf{D}\left(f - f\mathbf{H}_{(a)}\right) = \mathbf{D}f - \mathbf{D}\left(f\mathbf{H}_{(a)}\right)$$
$$= \mathbf{D}f - (\mathbf{D}f)\mathbf{H}_{(a)} - f(a)\delta_{(a)} = (\mathbf{D}f)\mathbf{H}_{(a)}^{*} - f(a)\delta_{(a)}.$$

Formulas (2.11) may be obtained by induction. The first one, for example, has already been proved for $n = 1$. Now suppose that it holds for $n = p$. Thus if the distribution $\mathbf{D}^{p}f$ is continuous at the point a, by applying the induction hypothesis to $g = \mathbf{D}f$ we get

$$\mathbf{D}^{p}\left(g\mathbf{H}_{(a)}\right) = \mathbf{D}^{p}g\mathbf{H}_{(a)} + \mathbf{D}^{p-1}g(a)\delta_{(a)} + \ldots + g(a)\delta_{(a)}^{(p-1)}.$$

Then, taking into account that

$$g\mathbf{H}_{(a)} = (\mathbf{D}f)\mathbf{H}_{(a)} = \mathbf{D}\left(f\mathbf{H}_{(a)}\right) - f(a)\delta_{(a)},$$

this gives

$$\mathbf{D}^{p}\left(\mathbf{D}f\mathbf{H}_{(a)} - f(a)\delta_{(a)}\right) = \left(\mathbf{D}^{p+1}f\right)\mathbf{H}_{(a)} + \mathbf{D}^{p}f(a)\delta_{(a)} + \ldots + \mathbf{D}f(a)\delta_{(a)}^{(p-1)}$$

or, again

$$\mathbf{D}^{p+1}\left(f\mathbf{H}_{(a)}\right) = \left(\mathbf{D}^{p+1}f\right)\mathbf{H}_{(a)} + \left(\mathbf{D}^{p}f(a)\right)\delta_{(a)} \ldots + f(a)\delta_{(a)}^{(p)}$$

as was to be shown. □

It should be remarked that the formulas (2.10) are consistent with the derivative of a "product" as obtained by the usual rule (for example: $\mathbf{D}(f\mathbf{H}_{(a)}) = (\mathbf{D}f)\mathbf{H}_{(a)} + f\delta_{(a)}$) provided we identify the "product" of the distribution f (continuous at the point a) by the Dirac distribution at the same point, with the distribution $f(a)\delta_{(a)}$ (that is, provided we set: $f\delta_{(a)} = f(a)\delta_{(a)}$). More generally, the formulas (2.11) are consistent with the derivative of order n of the "product" $f\mathbf{H}_{(a)}$ (or, $f\mathbf{H}_{(a)}^{*}$) if, supposing $\mathbf{D}^{n-1}f$ continuous at the point a, we attach to the "products" $\mathbf{D}^{k}f\delta_{(a)}^{(n-k-1)}$, $k = 0, \ldots, n-1$, a meaning equivalent to that considered in section 1.7 under the hypothesis that f was a function of class $\mathcal{C}^{\infty}$.

As an application we will see (in the last part of this section, and also in what follows) how the truncation may be used in the study of *the Cauchy problem* for a linear differential equation with constant coefficients, in the distributional setting.[7]

---

[7]The method of truncation is also applicable to the case of equations, or systems of equations, with variable coefficients, but, for the sake of greater simplicity, we will consider here only equations with constant coefficients. The method may also be applied to problems where the initial data has the form of limits or one-sided limits of the unknown distribution.

Consider the equation

$$a_0 \mathbf{D}^n u + a_1 \mathbf{D}^{n-1} u + \cdots + a_n u \;=\; f, \tag{2.12}$$

where $n \geq 1$, $a_0, a_1, \ldots, a_n$ are given complex constants ($a_0 \neq 0$), $u \in \mathcal{C}_\infty(\mathbb{R})$ is the unknown distribution and $f \in \mathcal{C}_\infty(\mathbb{R})$ is a given distribution. The values at the point 0 of the required distribution u and of its derivatives $\mathbf{D}u, \ldots, \mathbf{D}^{n-1}u$ are also given as follows:

$$u(0) = \alpha_0, \;\; \mathbf{D}u(0) = \alpha_1, \;\; \ldots, \mathbf{D}^{n-1}u(0) = \alpha_{n-1}, \tag{2.13}$$

with $\alpha_0, \alpha_1, \ldots, \alpha_{n-1} \in \mathbb{C}$.

We therefore ask not only that u satisfies (2.12) but that u and its derivatives of order not greater than $n-1$ should be continuous at the origin (and take at that point the given values); from this it follows that $\mathbf{D}^n u$ must be precontinuous at the point 0 (as is obviously the case with $u, \mathbf{D}u, \ldots, \mathbf{D}^{n-1}u$). Then, if the Cauchy problem has a solution, the right-hand side f of (2.12), which is a linear combination of distributions precontinuous at the point 0, must be a distribution precontinuous at the same point: hence the condition $f \in W_0(\mathbb{R})$ is necessary for a solution to the problem to exist.

We shall prove in the sequel that, reciprocally, if f is precontinuous at the origin, the Cauchy problem has in fact one and only one solution; and we shall also be able to construct an effective process to obtain that solution. For this purpose we will use, besides the truncation method, another important operation which will now be introduced only in a particular case but which will play a significant part in the sequel: this is the *convolution* of distributions.

The fundamental idea of the method of truncation, due to *O. Heaviside*, consists of the substitution of the truncated $u(t)\mathbf{H}(t)$ for the unknown $u(t)$. Intuitively, it might be thought that we are only interested in knowing the physical state of the system represented by the differential equation for positive values of the independent variable; if this variable represents time, this corresponds to the assumption that we are interested only in the "future". On the other hand, if we were interested only in the "past" it would be enough to use, instead of the truncation to the left, the truncation to the right of the origin, $u\mathbf{H}^*$).

To start with, suppose that there exists a solution u to the Cauchy problem (a hypothesis which must be confirmed later on) and apply to each one of the members of the equation:

$$a_0 \mathbf{D}^n u + a_1 \mathbf{D}^{n-1} u + \cdots + a_n u \;=\; f$$

the truncation to the left of the origin. This is legitimate since, if u is a solution of the problem, then all terms of the left-hand side will be, like f, distributions precontinuous at the point 0. We then obtain:

$$a_0 \left(\mathbf{D}^n u \right) \mathbf{H} + a_1 \left(\mathbf{D}^{n-1} u \right) \mathbf{H} + \cdots + a_n u \mathbf{H} \;=\; f \mathbf{H}. \tag{2.14}$$

Next, taking into account the initial conditions (2.13), the first formulas (2.11) give immediately:

$$
\begin{aligned}
\mathbf{D}(u\mathbf{H}) &= (\mathbf{D}u)\mathbf{H} + \alpha_0\delta \\
\mathbf{D}^2(u\mathbf{H}) &= (\mathbf{D}^2 u)\,\mathbf{H} + \alpha_1\delta + \alpha_0\delta' \\
&\;\;\vdots \\
\mathbf{D}^{n-1}(u\mathbf{H}) &= (\mathbf{D}^{n-1}u)\,\mathbf{H} + \alpha_{n-2}\delta + \alpha_{n-3}\delta' + \cdots + \alpha_0\delta^{(n-2)} \\
\mathbf{D}^n(u\mathbf{H}) &= (\mathbf{D}^n u)\,\mathbf{H} + \alpha_{n-1}\delta + \alpha_{n-2}\delta' + \cdots + \alpha_0\delta^{(n-1)}.
\end{aligned}
\tag{2.15}
$$

Substituting the values of $(\mathbf{D}^k u)\mathbf{H}$ given by the preceding formulae into (2.14), we get:

$$
a_0\mathbf{D}^n(u\mathbf{H}) + a_1\mathbf{D}^{n-1}(u\mathbf{H}) + \cdots + a_n u\mathbf{H} = f\mathbf{H} + \sum_{k=0}^{n-1}\beta_k\delta^{(k)},
\tag{2.16}
$$

where

$$
\beta_k = a_0\alpha_{n-k-1} + a_1\alpha_{n-k-2} + \cdots + a_{n-k-1}\alpha_0, \quad (k = 0, 1, \ldots, n-1).
$$

On the right-hand side of (2.16) there is a known distribution, null on $]-\infty, 0[$; denoting it by g, we may say that, if u is in fact a solution of the Cauchy problem, its truncation to the left of the origin, $u\mathbf{H}$, must be a solution of the following equation in w:

$$
a_0\mathbf{D}^n w + a_1\mathbf{D}^{n-1}w + \cdots + a_n w = g.
\tag{2.17}
$$

Clearly we may look at (2.17) as an equation in the space of all distributions which are null on $]-\infty, 0[$, a space which we will denote by the symbol $\mathcal{C}_\infty^+(\mathbb{R})$. This follows since $g \in \mathcal{C}_\infty^+(\mathbb{R})$ and the distribution w which we wish to obtain must also be an element of the same space.

In the next section we will prove that, whatever the right-hand side may be, (provided it belongs to $\mathcal{C}_\infty^+(\mathbb{R})$) an equation of the form (2.17) has one and only one solution in $\mathcal{C}_\infty^+(\mathbb{R})$.

It is easily seen that if we had started to truncate the two members of the equation (2.12) not to the left but to the right of the origin we would have similarly obtained:

$$
a_0\mathbf{D}^n(u\mathbf{H}^*) + a_1\mathbf{D}^{n-1}(u\mathbf{H}^*) + \cdots + a_n u\mathbf{H}^* = f\mathbf{H}^* - \sum_{k=0}^{n-1}\beta_k\delta^{(k)}.
$$

Hence, under the hypothesis that u is a solution of the Cauchy problem, we could guarantee that its truncation $u\mathbf{H}^*$ should be the solution of the equation in w^*:

$$
a_0\mathbf{D}^n w^* + a_1\mathbf{D}^{n-1}w^* + \cdots + a_n w^* = g^*
\tag{2.18}
$$

where $g^* = f\mathbf{H}^* - \sum_{k=0}^{n-1}\beta_k\delta^{(k)}$. This is an equation to be solved in $\mathcal{C}_\infty^-(\mathbb{R})$, the space of all distributions defined on $\mathbb{R}$ which are null on $]0, +\infty[$. This equation, as will be seen in the sequel, has a unique solution for any distribution which is null on the right-hand side.

Clearly the uniqueness of the solution of the equations (2.17) and (2.18), in the spaces $\mathcal{C}_\infty^+(\mathbb{R})$ and $\mathcal{C}_\infty^-(\mathbb{R})$, respectively, will guarantee the uniqueness of the solution of the Cauchy problem; for if $u\mathbf{H}$ and $u\mathbf{H}^*$ are uniquely determined then also is $u = u\mathbf{H} + u\mathbf{H}^*$. However the existence of the solutions w and w^* of the equations (2.17) and (2.18) is not enough to ensure the existence of a solution to the Cauchy problem, since the existence of that solution has been assumed from the outset. It remains to verify that the distribution $w + w^*$ is in fact a solution of the equation (2.12) and, moreover, satisfies the initial conditions.

The programme just sketched in general terms will be fully developed in the next section, making use of the fundamental concept of convolution introduced therein.

2.4 Convolution of Distributions which are Null to the Left of the Origin. Applications to Differential Equations.

Denote by $\mathcal{C}^+(\mathbb{R})$ (or simply by $\mathcal{C}^+$) the linear space of all complex-valued functions which are continuous on $\mathbb{R}$ and null on the interval $]-\infty, 0[$, and for each pair (F, G) of elements of $\mathcal{C}^+$ and each $x \in \mathbb{R}$, set

$$F * G(x) = \int_{-\infty}^{+\infty} F(x - t)G(t)\, dt. \tag{2.19}$$

We then have

> **Proposition 2.29** *Given F and G in $\mathcal{C}^+$, formula (2.19) defines a function $F * G : \mathbb{R} \to \mathbf{C}$ which is continuous on $\mathbb{R}$ and null on the left of the origin (that is, an element of $\mathcal{C}^+$).*

Proof: To start with note that since $G(t) = 0$ for $t < 0$ and $F(x - t) = 0$ for $t > x$ then formula (2.19) is equivalent to

$$F * G(x) = \int_0^x F(x - t)G(t)\, dt. \tag{2.20}$$

For any $x \in \mathbb{R}$ the integral on the right-hand side always exists since the integrand is continuous on the (compact) interval $[0, x]$. The value of that integral is precisely $F * G(x)$ and defines the function $F * G : \mathbb{R} \to \mathbf{C}$. If $x < 0$ then the integrand will be null over the whole interval and therefore we will have $F * G(x) = 0$.

To show that $F * G$ is a continuous function at every point $x \geq 0$ (and therefore on $\mathbb{R}$) suppose x fixed and, for any two arbitrarily chosen real numbers $a < 0$ and $b > x$, let $\mathbf{I} = [a, b]$, $\gamma = \max\{|a|, b - x\}$, $M = \sup_{t \in \mathbf{I}} |F(t)|$ and $N = \sup_{t \in \mathbf{I}} |G(t)|$.

Then, given $\delta > 0$, determine $\varepsilon > 0$ so that the two following conditions hold:

1) if $MN \neq 0$ then $\varepsilon < \delta/2MN$,

2) if $Nx \neq 0$, ε must be such that, whenever t and t^* are two points of $\mathbf{I}$ with $|t - t^*| < \varepsilon$, then:[8]

$$|F(t) - F(t^*)| < \frac{\delta}{2Nx}.$$

---

[8]Note that F is uniformly continuous on the interval $\mathbf{I}$.

Supposing $|h| < \varepsilon$ and $|h| < \gamma$, consider now the relations

$$|F * G(x+h) - F * G(x)| = \left| \int_0^{x+h} F(x-t+h)G(t)\, dt - \int_0^x F(x-t)G(t)\, dt \right|$$

$$\leq \int_0^x |F(x-t+h) - F(x-t)|\, |G(t)|\, dt$$

$$+ \left| \int_x^{x+h} |F(x-t+h)|\, |G(t)|\, dt \right|.$$

If $N = 0$ or $x = 0$, the first of these integrals will be null; if not we will have

$$\int_0^x |F(x-t+h) - F(x-t)|\, |G(t)|\, dt \ \leq\ \frac{\delta}{2Nx}\, Nx = \delta/2.$$

The second integral is zero if $M = 0$ or $N = 0$; if $MN \neq 0$ then

$$\int_x^{x+h} |F(x-t+h)|\, |G(t)|\, dt \ \leq\ \frac{\delta}{2MN}\, MN = \delta/2.$$

Hence, for $|h| < \min\{\varepsilon, \gamma\}$ we have

$$|F * G(x+h) - F * G(x)| \ <\ \delta,$$

thus proving that $F * G \in \mathcal{C}^+(\mathbb{R})$. $\qquad\qquad\square$

We have proved that $(F, G) \rightsquigarrow F * G$ is a binary operation on $\mathcal{C}^+$; it will be referred to as *convolution* or *convolution product*; the function $F * G$ is also usually called the *convolution* or the *product of convolution* of the functions F and G.

For each $m \in \mathbb{N}$ and $G \in \mathcal{C}^+$ denote by $\mathcal{J}^m G$ the primitive of order m of the function G that is again a function in $\mathcal{C}^+$. If $F \in \mathcal{C}^+$, the MacLaurin formula for $\mathcal{J}^{m+1} F$, with the integral form of remainder, may be written as

$$\mathcal{J}^{m+1} F(x) \ =\ \int_0^x \frac{(x-t)^m}{m!}\, F(t)\, dt.$$

It is then immediately obvious that $\mathcal{J}^{m+1} F$ is precisely the result of the convolution of F with the function $J_m(x) = \frac{x^m}{m!}\mathbf{H}(x)$:

$$J_m * F(x) \ =\ \int_0^x \frac{(x-t)^m}{m!}\, F(t)\, dt \ =\ \mathcal{J}^{m+1} F(x).$$

In particular, since $J_1 = J$ is the function which is null for $x \leq 0$ and equal to x for each point $x > 0$, we get:

$$J * F(x) \ =\ \mathcal{J}^2 F(x), \tag{2.21}$$

a formula which will be useful in the sequel.

Now for $\alpha \in \mathbb{C}$ and any positive integer m set:

$$J_m^\alpha(x) \ =\ \frac{x^m}{m!}\, e^{\alpha x} \mathbf{H}(x).$$

For any positive integer n we will have:

$$J_m^\alpha * J_n^\alpha(x) = \left(\int_0^x \frac{(x-t)^m}{m!} e^{\alpha(x-t)} \frac{t^n}{n!} e^{\alpha t} \, dt \right) \mathbf{H}(x)$$

$$= \left(\int_0^x \frac{(x-t)^m}{m!} \frac{t^n}{n!} \, dt \right) e^{\alpha x} \mathbf{H}(x)$$

$$= \frac{x^{m+n+1}}{(m+n+1)!} e^{\alpha x} \mathbf{H}(x) = J_{m+n+1}^\alpha(x),$$

that is

$$J_m^\alpha * J_n^\alpha = J_{m+n+1}^\alpha. \tag{2.22}$$

Next we have the result:

Proposition 2.30 *The convolution is an operation which is commutative, associative and distributive with respect to the addition of functions.*

Proof: With respect to commutativity it is enough to make the change of variable $t = x - u$ on the integral of the right-hand side of (2.20) to obtain:

$$F * G(x) = \int_0^x F(u)G(x-u) \, du = G * F(x).$$

For the distributivity with respect to addition it is enough to note that

$$F * (G + H)(x) = \int_0^x F(x-t)[G(t) + H(t)] \, dt$$

$$= \int_0^x F(x-t)G(t) \, dt + \int_0^x F(x-t)H(t) \, dt$$

$$= F * G(x) + F * H(x).$$

Finally to prove associativity we remark that, denoting by $\mathbf{A}$ the subset of $\mathbb{R}^2 = \mathbb{R}_u \times \mathbb{R}_v$ defined by the conditions

$$u \geq 0, \quad v \geq 0, \quad v \leq x \text{ and } u \leq v,$$

we have

$$(F * G) * H(x) = \int_0^x \left(\int_0^v F(v-u)G(u) \, du \right) H(x-v) \, dv$$

$$= \int\!\!\int_{\mathbf{A}} F(v-u)G(u)H(x-v) \, du \, dv.$$

In the last integral, the change of variables $(u, v) \rightsquigarrow (\xi, \eta)$ defined by

$$\begin{cases} u = x - \xi - \eta \\ v = x - \eta, \end{cases}$$

transforms $\mathbf{A}$ into the subset $\mathbf{A}^*$ of $\mathbb{R}_\xi \times \mathbb{R}_\eta$ defined by

$$\xi \geq 0, \quad \eta \geq 0, \text{ and } \xi + \eta \leq x.$$

Then, noting that the Jacobian of the transformation is equal to 1, we obtain:

$$
\begin{aligned}
(F * G) * H(x) &= \iint_{\mathbf{A}^*} F(\xi)G(x - \xi - \eta)H(\eta)\, d\xi d\eta \\
&= \int_0^x F(\xi)\left(\int_0^{x-\xi} G(x - \xi - \eta)H(\eta)\, d\eta\right) d\xi \\
&= F * (G * H)(x),
\end{aligned}
$$

as was to be proved. $\qquad\square$

In order to be able to extend the notion of convolution from $\mathcal{C}^+$ to the space $\mathcal{C}^+_\infty(\mathbb{R})$, of distributions defined on $\mathbb{R}$ and null to the left of the origin, we begin with the following proposition:

> **Proposition 2.31** *Let F and G be any two functions in $\mathcal{C}^+$; if $F \in \mathcal{C}^1(\mathbb{R})$ then $F * G$ is a function in $\mathcal{C}^1(\mathbb{R})$ also and, moreover, $(F * G)' = F' * G$.*

Proof: Over the interval $] - \infty, 0[$ we have $F * G = 0$ and $F' * G = 0$ and therefore the equality holds. For $x \geq 0$ set, as in the proof of 2.29, $\mathbf{I} = [a, b]$ with $a < 0 \leq x < b$, $\gamma = \max\{|a|, b - x\}$, $M = \sup_{t \in \mathbf{I}} |F(t)|$ and $N = \sup_{t \in \mathbf{I}} |G(t)|$. Taking into account that under the conditions of the hypothesis we have

$$
F(x - t + h) - F(x - t) = hF'(x - t + \theta_1 h)
$$

and, (from the mean value theorem),

$$
\int_x^{x+h} F(x - t + h)G(t)\, dt = hF(h(1 - \theta_2))G(x + \theta_2 h)
$$

with $\theta_1, \theta_2 \in]0, 1[$, we get:

$$
\begin{aligned}
F * G(x + h) - F * G(x) &= \int_0^x [F(x - t + h) - F(x - t)]G(t)\, dt \\
&\quad + \int_x^{x+h} F(x - t + h)G(t)\, dt \\
&= h\int_0^x F'(x - t + \theta_1 h)G(t)\, dt \\
&\quad + hF(h(1 - \theta_2))G(x + \theta_2 h).
\end{aligned}
$$

Given $\delta > 0$, let $\varepsilon > 0$ be chosen so that

(1) if $N \neq 0$, then $|F(t)| < \delta/2N$ if $|t| < \varepsilon$;

(2) if $Nx \neq 0$, we have

$$
|F'(t) - F'(t^*)| < \frac{\delta}{2Nx},
$$

whenever $t, t^* \in \mathbf{I}$ and $|t - t^*| < \varepsilon$.

Then, for $|h| < \min\{\varepsilon, \gamma\}$, we will have

$$\left| \frac{1}{h}[F * G(x + h) - F * G(x)] - F' * G(x) \right| \leq$$

$$\left| \int_0^x [F'(x - t + \theta_1 h) - F'(x - t)] G(t) \, dt \right| + |F(h(1 - \theta_2))G(x + \theta_2 h)|.$$

The second term of the second member will be null for $N = 0$ and, if $N \neq 0$ by (1) we will have:

$$|F(h(1 - \theta_2))G(x + \theta_2 h)| < \frac{\delta}{2N} \; N = \delta/2.$$

The first term will be null if $x = 0$ or $N = 0$. For the other cases we will have:

$$\left| \int_0^x [F'(x - t + \theta_1 h) - F'(x - t)] G(t) \, dt \right| \leq xN \frac{\delta}{2Nx} = \delta/2.$$

It may therefore be concluded that, for $|h| < \min\{\varepsilon, \gamma\}$, we have

$$\left| \frac{1}{h}[F * G(x + h) - F * G(x)] - F' * G(x) \right| < \delta,$$

that is, $F * G$ is differentiable on $\mathbb{R}$, $(F * G)' = F' * G$ being an element of $\mathcal{C}^+$. $\square$

From this result, together with the commutativity of the convolution, it follows immediately that:

Corollary 2.32 *Let F and G be two functions in $\mathcal{C}^+(\mathbb{R})$.*

(1) *If $F \in \mathcal{C}^p(\mathbb{R})$ and $G \in \mathcal{C}^q(\mathbb{R})$ then $F * G \in \mathcal{C}^{p+q}(\mathbb{R})$ and*

$$(F * G)^{(p+q)} = F^{(p)} * G^{(q)}.$$

(2) *If $F \in \mathcal{C}^\infty(\mathbb{R})$ then $F * G \in \mathcal{C}^\infty(\mathbb{R})$ and, for any $p \in \mathbb{N}$:*

$$(F * G)^{(p)} = F^{(p)} * G.$$

The first part of this corollary suggests the way that we shall adopt in order to extend the operation of convolution to distributions which are null to the left of the origin; however, before we give the definition, it is convenient to state the following proposition:

Proposition 2.33 *If $f \in \mathcal{C}^+_\infty(\mathbb{R})$ then there exist $n \in \mathbb{N}$ and $F \in \mathcal{C}^+(\mathbb{R})$ such that $f = \mathbf{D}^n F$.*

Proof: Supposing that $f = \mathbf{D}^n \Phi$, with $\Phi \in \mathcal{C}(\mathbb{R})$, the restriction of Φ to $]-\infty, 0[$ must coincide with the restriction to the same interval of a polynomial P, of degree $< n$; thus we also have $f = \mathbf{D}^n F$ with $F = \Phi - P$, and clearly $F \in \mathcal{C}^+(\mathbb{R})$. $\square$

Proposition 2.34 *There exists one and only one map $(f, g) \rightsquigarrow f * g$, from $\mathcal{C}^+_\infty \times \mathcal{C}^+_\infty$ into $\mathcal{C}^+_\infty$ such that*

c1 *If $f, g \in C^+$ then $f * g$ is the convolution of the functions f and g, as defined above;*

c2 *for every $f, g \in C_\infty^+$, we have $\mathbf{D}(f * g) = (\mathbf{D}f) * g = f * (\mathbf{D}g)$.*

Proof: The uniqueness is immediate: a pair of distributions (f, g) which are null at the left of the origin, may be represented in the form

$$f = \mathbf{D}^m F, \quad g = \mathbf{D}^n G,$$

with $F, G \in C^+$ and $m, n \in \mathbb{N}$. Then (if the map in question exists), by **c1** and **c2**, we will have necessarily:

$$f * g = \mathbf{D}^m F * \mathbf{D}^n G = \mathbf{D}^{m+n}(F * G).$$

To prove existence suppose we have both $f = \mathbf{D}^m F$ and $f = \mathbf{D}^p \Phi$, where F and Φ belong to C^+. If, for example, $p \geq m$ then since both F and Φ are null on $]-\infty, 0[$, we should have:

$$\mathcal{J}^m \Phi - \mathcal{J}^p F = 0$$

and thus

$$F = \Phi^{(p-m)}.$$

From this it follows that

$$F * G = \Phi^{(p-m)} * G = (\Phi * G)^{(p-m)}$$

and

$$\mathbf{D}^{m+n}(F * G) = \mathbf{D}^{m+n}(\Phi * G)^{(p-m)} = \mathbf{D}^{p+n}(\Phi * G).$$

It is clear that if we also have $g = \mathbf{D}^q \Psi$ with $\Psi \in C^+$, then similarly we would obtain the equality:

$$\mathbf{D}^{p+n}(\Phi * G) = \mathbf{D}^{p+q}(\Phi * \Psi)$$

and therefore

$$\mathbf{D}^{m+n}(F * G) = \mathbf{D}^{p+q}(\Phi * \Psi),$$

so that the proof is complete. $\qquad\square$

We will naturally refer to the distribution $f * g$ as the *convolution* or *convolution product* of f and g. We may now state the following proposition:

Proposition 2.35 *The set C_∞^+, with its usual linear space structure and with the convolution operation, forms a complex commutative algebra with unit element.*

Proof: That the convolution is an operation which is associative, commutative and distributive with respect to addition is an immediate consequence of the definition of convolution in C_∞^+ and of the proposition 2.30. For the associativity, for example, if we have $f = \mathbf{D}^m F$, $g = \mathbf{D}^n G$ and $h = \mathbf{D}^p H$ with $m, n, p \in \mathbb{N}$ and $F, G, H \in C^+$, we will obtain:

$$\begin{aligned}
(f * g) * h &= [\mathbf{D}^{m+n}(F * G)] * \mathbf{D}^p H \\
&= \mathbf{D}^{m+n+p}[(F * G) * H] = \mathbf{D}^{m+n+p}[F * (G * H)] \\
&= \mathbf{D}^m F * \mathbf{D}^{n+p}(G * H) = f * (g * h).
\end{aligned}$$

The property

$$\alpha(f * g) \;=\; (\alpha f) * g \;=\; f * (\alpha g)$$

which is also easily proved for $\alpha \in \mathbf{C}$ and $f, g \in \mathcal{C}_\infty^+$, allows us to conclude that $\mathcal{C}_\infty^+$ is a *commutative* algebra.

Finally, if f is an arbitrary element in $\mathcal{C}_\infty^+$, so that $f = \mathbf{D}^n F$ with $F \in \mathcal{C}^+$, then using (2.21), we will have:

$$f * \delta \;=\; \mathbf{D}^n F * \mathbf{D}^2 J \;=\; \mathbf{D}^{n+2}(F * J) \;=\; \mathbf{D}^{n+2} \mathcal{J}^2 F \;=\; \mathbf{D}^n F \;=\; f.$$

The algebra $\mathcal{C}_\infty^+$ therefore has a unit element, namely the distribution δ. $\qquad\qquad\square$

Clearly, if $f = \mathbf{D}^m F$ and $g = \mathbf{D}^n G$ with $F, G \in \mathcal{C}^+$, we will have:

$$\mathbf{D}(f * g) \;=\; (\mathbf{D}f) * g \;=\; f * (\mathbf{D}g) \;=\; \mathbf{D}^{m+n+1}(F * G).$$

Hence, also in the space of distributions which are null to the left of the origin, *to obtain the derivative of the convolution of two distributions it is enough to convolve one of the distributions with the derivative of the other.* From this it follows at once that:

$$f * \delta' \;=\; \mathbf{D}(f * \delta) \;=\; \mathbf{D}f$$

and, more generally,

$$f * \delta^{(p)} \;=\; \mathbf{D}^p f. \tag{2.23}$$

In this way a linear differential equation with constant coefficients

$$a_0 \mathbf{D}^n w + a_1 \mathbf{D}^{n-1} w + \cdots + a_n w \;=\; g \quad (a_0 \neq 0), \tag{2.24}$$

(with the unknown w and the distribution g in $\mathcal{C}_\infty^+$) may be written as a *convolution equation*,[9] in the form:

$$a * w \;=\; g,$$

where

$$a \;=\; a_0 \delta^{(n)} + a_1 \delta^{(n-1)} + \cdots + a_n \delta.$$

It is easy to see that:

Proposition 2.36 *Let $a \in \mathcal{C}_\infty^+(\mathbb{R})$. For the convolution equation*

$$a * w \;=\; g, \tag{2.25}$$

(where g is a given distribution in $\mathcal{C}_\infty^+$ and w, the unknown distribution, is supposed to belong to the same space) to have at least one solution for any g, it is necessary and sufficient that a is an invertible element of the algebra $\mathcal{C}_\infty^+$; that is, that there exists $a^{-1} \in \mathcal{C}_\infty^+$ such that

$$a * a^{-1} \;=\; a^{-1} * a \;=\; \delta.$$

If this condition is satisfied then, for each $g \in \mathcal{C}_\infty^+$, the unique solution of (2.25) is the distribution $w = a^{-1} * g$.

---

[9]Several important types of integral, differential, integro-differential, delay-differential, etc. equations may also assume the form of convolution equations, which contributes to the great importance of this kind of equation.

Proof: If equation (2.25) has a solution for any right-hand side g, it must also have a solution in particular for $g = \delta$, that is, a must be invertible. Conversely, if a is invertible, $a^{-1} * g$ is a solution of the equation (2.25) since:

$$a * \left(a^{-1} * g\right) = \left(a * a^{-1}\right) * g = \delta * g = g.$$

Moreover, if we have $a * \nu = g$, then we also have:

$$\nu = \left(a^{-1} * a\right) * \nu = a^{-1} * (a * \nu) = a^{-1} * g,$$

which proves the uniqueness of the solution. $\qquad\square$

We remark that the *preceding proposition* 2.36, (stated and proved here by considering directly the algebra $C_\infty^+(\mathbb{R})$ and its unit, δ) *remains valid, with a completely similar proof, if $C_\infty^+(\mathbb{R})$ is replaced by an arbitrary algebra with unit element.*

We will prove now that any element in $C_\infty^+(\mathbb{R})$ of the form

$$a = a_0 \delta^{(n)} + a_1 \delta^{(n-1)} + \cdots + a_n \delta,$$

with $a_0 \neq 0$, is invertible, thereby guaranteeing the existence and uniqueness of the solution to the convolution equation $a * w = g$, or of the corresponding differential equation (2.24). For this purpose, consider the homogeneous equation associated with (2.24)

$$a_0 \varphi^{(n)} + a_1 \varphi^{(n-1)} + \cdots + a_n \varphi = 0, \tag{2.26}$$

where we have denoted the unknown by φ.

In the classical setting of the theory of functions of a real variable, the existence and uniqueness theorem guarantees that there exists one and only one function θ which is a solution of this equation and which satisfies (in the usual sense) the following initial conditions:

$$\theta(0) = \theta'(0) = \cdots = \theta^{(n-2)}(0) = 0, \quad \theta^{(n-1)}(0) = \frac{1}{a_0}.$$

Clearly θ, as with any other function which is a solution of a linear differential equation with constant coefficients, is a function in C^∞. Setting $\omega = \theta \mathbf{H}$ we will have

$$
\begin{aligned}
\mathbf{D}\omega &= (\mathbf{D}\theta)\mathbf{H} + \theta(0)\delta = (\mathbf{D}\theta)\mathbf{H} \\
\mathbf{D}^2\omega &= (\mathbf{D}^2\theta)\mathbf{H} + \theta'(0)\delta + \theta(0)\delta' = (\mathbf{D}^2\theta)\mathbf{H} \\
&\;\;\vdots \\
\mathbf{D}^{n-1}\omega &= (\mathbf{D}^{n-1}\theta)\mathbf{H} + \theta^{(n-2)}(0)\delta + \cdots + \theta(0)\delta^{(n-2)} = (\mathbf{D}^{n-1}\theta)\mathbf{H} \\
\mathbf{D}^n\omega &= (\mathbf{D}^n\theta)\mathbf{H} + \theta^{(n-1)}(0)\delta + \cdots + \theta(0)\delta^{(n-1)} = (\mathbf{D}^n\theta)\mathbf{H} + \tfrac{1}{a_0}\delta
\end{aligned}
$$

and therefore, since θ is a solution of the homogeneous equation (2.26):

$$a_0 \mathbf{D}^n\omega + a_1 \mathbf{D}^{n-1}\omega + \cdots + a_n \omega = 0 \cdot \mathbf{H} + a_0 \frac{1}{a_0}\delta = \delta.$$

This equation may be written in terms of convolution as:

$$\left(a_0 \delta^{(n)} + a_1 \delta^{(n-1)} + \cdots + a_n \delta\right) * \omega = \delta$$

which proves the invertibility of the element $a_0 \delta^{(n)} + a_1 \delta^{(n-1)} + \cdots + a_n \delta$ in the algebra C_∞^+. We may therefore state:

Proposition 2.37 *If n is an integer ≥ 1, $a_0, a_1, \ldots, a_n \in \mathbf{C}$, and $a_0 \neq 0$, then the differential equation:*

$$a_0 \mathbf{D}^n w + a_1 \mathbf{D}^{n-1} w + \cdots + a_n w = g \qquad (2.27)$$

has, for each $g \in \mathcal{C}_\infty^+(\mathbb{R})$, one and only one solution in the same space.

The distribution ω just obtained is the unique solution, null to the left of the origin, of the equation in the unknown w:

$$a_0 \mathbf{D}^n w + a_1 \mathbf{D}^{n-1} w + \cdots + a_n w = \delta$$

and is usually called the *elementary solution* of the equation (2.27), or of the differential operator $a_0 \mathbf{D}^n + a_1 \mathbf{D}^{n-1} + \cdots + a_n$. When this solution is known, the solution of the nonhomogeneous equation may be obtained by a simple convolution.

We now consider some examples of elementary solutions in the convolution algebra $\mathcal{C}_\infty^+(\mathbb{R})$.

1. For the first order linear equation

$$(\mathbf{D} - \alpha) w = g,$$

with $\alpha \in \mathbf{C}$, the solution of the associated homogeneous equation that has the value 1 at the point 0 is $e^{\alpha x}$; the elementary solution is therefore $e^{\alpha x} \mathbf{H}(x)$.

2. For the second order equation reducible to the form

$$(\mathbf{D} - \alpha)(\mathbf{D} - \beta) w = g,$$

where α and β are the roots of the corresponding characteristic polynomial, the elementary solution can also be easily obtained. This equation is equivalent to the convolution equation

$$[(\delta' - \alpha\delta) * (\delta' - \beta\delta)] * w = g,$$

and the elementary solution, which is the inverse of the distribution

$$(\delta' - \alpha\delta) * (\delta' - \beta\delta),$$

may be obtained by convolving the inverses of $\delta' - \alpha\delta$ and $\delta' - \beta\delta$, that is the distributions $e^{\alpha x} \mathbf{H}(x)$ and $e^{\beta x} \mathbf{H}(x)$. Simple calculations lead to

$$e^{\alpha x} \mathbf{H}(x) * e^{\beta x} \mathbf{H}(x) = \begin{cases} \frac{e^{\alpha x} - e^{\beta x}}{\alpha - \beta} \mathbf{H}(x) & \text{if } \alpha \neq \beta \\ x e^{\alpha x} \mathbf{H}(x) & \text{if } \alpha = \beta. \end{cases}$$

In particular, if α and β are complex conjugate, $\alpha = r + is$, $\beta = r - is$, with r, s real and $s \neq 0$, the elementary solution may assume the form:

$$\frac{e^{(r+is)x} - e^{(r-is)x}}{2is} \mathbf{H}(x) = e^{rx} \frac{\sin(sx)}{s} \mathbf{H}(x).$$

3. The process described in **2.** may obviously be generalised. For an operator of order n, for which the roots $\alpha_1, \alpha_2, \ldots, \alpha_n$ of its characteristic polynomial are known, the elementary solution will be

$$e^{\alpha_1 x}\mathbf{H}(x) * e^{\alpha_2 x}\mathbf{H}(x) * \cdots * e^{\alpha_n x}\mathbf{H}(x).$$

In the particular case that $\alpha_1 = \alpha_2 = \cdots = \alpha_n = \alpha$ it is easily seen that

$$(\delta' - \alpha\delta)^{-n} = \frac{x^{n-1}}{(n-1)!}\, e^{\alpha x}\mathbf{H}(x).$$

We remark that everything we have seen with respect to the convolution algebras $\mathcal{C}^+(\mathbb{R})$ and $\mathcal{C}^+_\infty(\mathbb{R})$ could be repeated, with slight modifications, for $\mathcal{C}^-(\mathbb{R})$, the space of continuous functions which are null for $x > 0$, and $\mathcal{C}^-_\infty(\mathbb{R})$, the space of distributions which are null to the right of the origin. In $\mathcal{C}^-$ the convolution is defined by the formula:

$$F * G(x) = \int_{-\infty}^{+\infty} F(x-t)G(t)\, dt = \int_x^0 F(x-t)G(t)\, dt$$

the extension to $\mathcal{C}^-_\infty$ being given by

$$f * g = \mathbf{D}^{m+n}(F * G),$$

for $f = \mathbf{D}^m F$ and $g = \mathbf{D}^n G$ with $F, G \in \mathcal{C}^-$.

Similarly we could prove that with its usual linear space structure, $\mathcal{C}^-_\infty$ is a convolution algebra with δ as its unit element; in this algebra any element of the form

$$a = a_0\delta^{(n)} + a_1\delta^{(n-1)} + \cdots + a_n\delta \quad (a_0 \neq 0)$$

is invertible and has the distribution $-\theta\mathbf{H}^*$ as its inverse, where θ still denotes the function that satisfies the conditions:

$$a_0\theta^{(n)} + a_1\theta^{(n-1)} + \cdots + a_n\theta = 0$$

$$\theta(0) = \theta'(0) = \cdots = \theta^{(n-2)}(0) = 0, \quad \theta^{(n-1)}(0) = \tfrac{1}{a_0}.$$

It follows therefore that in $\mathcal{C}^-_\infty(\mathbb{R})$, any convolution equation

$$a * w^* = g^*$$

has one and only one solution.

Thus if the right-hand member, f, of the equation (2.12) is a distribution which is precontinuous at the origin, the existence of unique solutions to the equations (2.17) and (2.18) has been proved and, at the same time, so has the uniqueness of the Cauchy problem. Denoting by w and w^*, respectively, the solutions of (2.17) and (2.18), we now try to prove that the distribution $u = w + w^*$ is, in fact, a solution of that problem.

First, adding equations (2.17) and (2.18) gives

$$a_0\mathbf{D}^n u + a_1\mathbf{D}^{n-1}u + \cdots + a_n u = g + g^* = f\mathbf{H} + f\mathbf{H}^* = f,$$

and this shows that u is a solution of the equation (2.12). To see that this solution also satisfies the initial conditions (2.13) we begin by proving the following lemma:

Lemma 2.38 *If the equality*

$$a_0 \mathbf{D}^n v + a_1 \mathbf{D}^{n-1} v + \cdots + a_n v = f \quad (a_0 \neq 0)$$

holds and if f is a distribution precontinuous at the point 0, the distribution $\mathbf{D}^n v$ will also be precontinuous at the same point.

Proof: For each integer $p \geq 0$ set

$$\begin{aligned}
V^{(p)} &= \mathbf{D}^p[V_0(\mathbb{R})] &= \{\mathbf{D}^p h : h \in V_0(\mathbb{R})\} \\
V^{(-p)} &= \mathbf{D}^{-p}[V_0(\mathbb{R})] &= \{h \in \mathcal{C}_\infty(\mathbb{R}) : \mathbf{D}^p h \in V_0(\mathbb{R})\},
\end{aligned}$$

where $V_0(\mathbb{R})$ denotes as usual the space of distributions defined on $\mathbb{R}$ and continuous at the point 0. We will have, in particular, $V^{(0)} = V_0(\mathbb{R})$, $V^{(1)}(\mathbb{R}) = W_0(\mathbb{R})$ and from the relation $V^{(0)} \subset V^{(1)}$ it follows at once that $V^{(p)} \subset V^{(p+1)}$ for each $p \in \mathbb{Z}$.

On the other hand, since any distribution defined on $\mathbb{R}$ is the derivative of some finite order of a continuous function (and therefore an element of $V_0(\mathbb{R})$), it is easy to see that

$$\bigcup_{p \in \mathbb{Z}} V^{(p)}(\mathbb{R}) = \mathcal{C}_\infty(\mathbb{R}).$$

Suppose now that $\nu \in \mathcal{C}_\infty(\mathbb{R})$ is a distribution such that $\mathbf{D}^n \nu \notin V^{(1)}(\mathbb{R}) = W_0(\mathbb{R})$. It is immediately seen that there must exist $p \in \mathbb{Z}$ such that $\nu \in V^{(p)} \backslash V^{(p-1)}$ and therefore that

$$\begin{aligned}
\mathbf{D}\nu &\in V^{(p+1)} \backslash V^{(p)} \\
&\;\;\vdots \\
\mathbf{D}^{n-1}\nu &\in V^{(p+n-1)} \backslash V^{(p+n-2)} \\
\mathbf{D}^n \nu &\in V^{(p+n)} \backslash V^{(p+n-1)}.
\end{aligned} \tag{2.28}$$

Since $\mathbf{D}^n \nu \notin V^{(1)}$ and $\mathbf{D}^n \nu \in V^{(p+n)}$ we must have $p + n > 1$.

On the other hand, the first $n - 1$ relations (2.28) allow us to see that each one of the distributions $\nu, \mathbf{D}\nu, \mathbf{D}^2\nu, \ldots, \mathbf{D}^{n-1}\nu$ belongs to the space $V^{(p+n-1)}$ while the last of those relations shows that $\mathbf{D}^n \nu \notin V^{(p+n-1)}$. Hence, if

$$f = a_0 \mathbf{D}^n \nu + a_1 \mathbf{D}^{n-1} \nu + \cdots + a_n \nu$$

we cannot have $f \in V^{(p+n-1)}$; since $V^{(1)} \subset V^{(p+n-1)}$ for $p + n > 1$ it follows that the distribution f cannot be precontinuous at the point 0. $\qquad\square$

The preceding lemma shows that for the solution $u = w + w^*$ of the equation (2.12) we have $\mathbf{D}^n u \in W_0(\mathbb{R})$ (since f is supposed to be precontinuous at the origin); it then follows that the distributions $\mathbf{D}^{n-1}u, \ldots, \mathbf{D}u$ and u are continuous at the point 0. Once continuity has been established, it will be enough to see that

$$u(0^+) = \alpha_0, \quad \mathbf{D}u(0^+) = \alpha_1, \quad \ldots, \quad \mathbf{D}^{n-1}u(0^+) = \alpha_{n-1}.$$

in order to prove that the initial conditions (2.13) are satisfied. From now on we adopt the convention that, for every $k \in \mathbb{N}$ and every $h \in \mathcal{C}_\infty^+$, $\mathbf{D}^{-k}h$ denotes the unique distribution which is null to the left of the origin and for which the k-th order derivative is equal to h. Now recall the equation (2.17) written in the form

$$a_0 \mathbf{D}^n w + a_1 \mathbf{D}^{n-1} w + \cdots + a_n w = f\mathbf{H} + \sum_{k=0}^{n-1} \beta_k \delta^{(k)},$$

Applying to both members of this equation the linear operator $\mathbf{D}^{-n}$, we get

$$a_0 w + a_1 \mathbf{D}^{-1} w + \cdots + a_n \mathbf{D}^{-n} w = \mathbf{D}^{-n}(f\mathbf{H}) + \beta_{n-1}\mathbf{H} + \sum_{k=0}^{n-2} \beta_k \mathbf{D}^{-n} \delta^{(k)}. \qquad (2.29)$$

The right-hand side of this equation is a distribution precontinuous at the point 0 (all terms of the right-hand side are continuous except $\beta_{n-1}\mathbf{H}(x)$ which is a distribution precontinuous at the origin); from the lemma 2.38 it follows that $w = \mathbf{D}^n(\mathbf{D}^{-n}w)$ is a distribution which is precontinuous at the origin. The distributions $\mathbf{D}^{-1}w, \mathbf{D}^{-2}w, \ldots, \mathbf{D}^{-n}w$ will therefore all be continuous at the point 0 at which, because they have limit zero to the left, they will have the value 0; obviously the same can be said with respect to the distributions $\mathbf{D}^{-k}(f\mathbf{H})$ for $k = 1, \ldots, n$ and $\mathbf{D}^{-k}\delta^{(j)}$ if $k > j + 1$.

Under these conditions passing to the limit when $x \to 0^+$ in the equation (2.29) we will obtain

$$a_0 w(0^+) = \beta_{n-1} = a_0 \alpha_0$$

and therefore

$$u(0) = u(0^+) = w(0^+) + w^*(0^+) = \alpha_0.$$

Differentiating (2.29) and again making $x \to 0^+$, we get:

$$a_0 \mathbf{D}w + a_1 w + \quad \cdots \quad + a_n \mathbf{D}^{-n+1} w = \mathbf{D}^{-n+1}(f\mathbf{H})$$
$$+ \sum_{k=0}^{n-3} \beta_k \mathbf{D}^{-n+1} \delta^{(k)} + \beta_{n-2}\mathbf{H} + \beta_{n-1}\delta$$

and

$$a_0 \mathbf{D}w(0^+) + a_1 \alpha_0 = \beta_{n-2} = a_0 \alpha_1 + a_1 \alpha_0$$

and therefore

$$\mathbf{D}u(0) = \mathbf{D}w(0^+) = \alpha_1.$$

In general, supposing that, for $k = 0, 1, \ldots, p-1 \, (p < n)$, the equalities $\mathbf{D}^k u(0) = \alpha_k$ have been proved, it will be enough to consider the relation

$$a_0 \mathbf{D}^p w + a_1 \mathbf{D}^{p-1} w + \cdots + a_n \mathbf{D}^{-n+p} w =$$

$$\sum_{k=0}^{n-p-2} \beta_k \mathbf{D}^{-n+p} \delta^{(k)} + \beta_{n-p-1}\mathbf{H} + \beta_{n-p}\delta + \cdots + \beta_{n-1}\delta^{(p-1)},$$

obtained from (2.29) by applying the operator $\mathbf{D}^p$ to both members and then making $x \to 0^+$ to obtain

$$a_0\mathbf{D}^p w(0^+) + a_1\alpha_{p-1} + \cdots + a_p\alpha_0 \;=\; \beta_{n-p-1} \;=\; a_0\alpha_p + a_1\alpha_{p-1} + \cdots + a_p\alpha_0.$$

Hence we conclude that

$$\mathbf{D}^p u(0) \;=\; \mathbf{D}^p w(0^+) \;=\; \alpha_p.$$

We may therefore state:

> **Theorem 2.39** *Let $a_0, a_1, \ldots, a_n$ and $\alpha_0, \alpha_1, \ldots, \alpha_{n-1}$ be complex numbers with $a_0 \neq 0$. A necessary and sufficient condition for the equation*
>
> $$a_0\mathbf{D}^n u + a_1\mathbf{D}^{n-1}u + \cdots + a_n u \;=\; f$$
>
> *to have a solution u satisfying the initial conditions*
>
> $$u(0) = \alpha_0, \quad \mathbf{D}u(0) = \alpha_1, \quad \ldots, \quad \mathbf{D}^{n-1}u(0) = \alpha_{n-1}$$
>
> *is that the distribution f be precontinuous at the point 0. If this condition holds then the solution is unique.*

2.5   Orders of Growth of Distributions. Integration.

The notations "**O**" and "**o**" of Landau are very useful in classical Analysis. They may be introduced according to the following definitions (in which we take as a model the case $x \to +\infty$, the adaptations to be made in the cases $x \to -\infty$, $x \to a$, $x \to a^+$, etc. with $a \in \mathbb{R}$ being easy to see).

Let F and φ be functions defined on intervals of $\mathbb{R}$ which are unbounded to the right. We say that *F is negligible with respect to φ when $x \to +\infty$* (or to the right) if there exists an interval $\mathbf{I}$ unbounded to the right and contained in the domains of F and φ, and a function F_0 defined on $\mathbf{I}$ and converging to 0 when $x \to +\infty$ such that $F = \varphi F_0$ on $\mathbf{I}$. On the other hand, we say that *F is dominated by φ when $x \to +\infty$* if there exists an interval $\mathbf{I}$, unbounded to the right, and a function F_0 defined and bounded on $\mathbf{I}$ such that $F = \varphi F_0$ on $\mathbf{I}$.

In the first case we write:[10] $F \in \mathbf{o}(\varphi)$ when $x \to +\infty$, and in the second case $F \in \mathbf{O}(\varphi)$ when $x \to +\infty$, the expression "when $x \to +\infty$" being very often omitted when no confusion should arise.

Clearly if there exists an interval, unbounded to the right, on which the function φ is never zero, we have $F \in \mathbf{o}(\varphi)$ if and only if $\lim_{x \to +\infty} F(x)/\varphi(x) = 0$, and $F \in \mathbf{O}(\varphi)$ if and only if F/φ is a function bounded on a neighbourhood of $+\infty$.

If φ is a function in C^∞ the symbol "$\mathbf{o}$" may be extended immediately to apply to distributions, in accordance with the following: A distribution f is said to be

---

[10] We prefer this, more logically correct, notation to the classical $F = \mathbf{o}(\varphi)$, $F = \mathbf{O}(\varphi)$.

negligible with respect to the function φ (both defined on intervals unbounded to the right), and we write $f \in \mathbf{o}(\varphi)$ if there exists an interval $\mathbf{I}$, unbounded to the right, and a distribution $f_0 \in \mathcal{C}_\infty(\mathbf{I})$ such that $\lim_{x \to +\infty} f_0(x) = 0$ and $f = \varphi f_0$ on $\mathbf{I}$.

For example, if $\alpha \in \mathbb{R}$ we will have $f \in \mathbf{o}(x^\alpha)$ if there exists a distribution f_0 converging to 0 when $x \to +\infty$ and such that $f(x) = x^\alpha f_0(x)$ on an interval which is unbounded to the right; in particular the condition $f \in \mathbf{o}(1)$ is equivalent to the statement $\lim_{x \to +\infty} f(x) = 0$.

Before defining the symbol $\mathbf{O}$ in a distributional setting it is convenient to make a brief reference to the concept of *bounded distribution*.

Let $\mathbf{I}$ be an interval which is unbounded to the right and let $f \in \mathcal{C}_\infty(\mathbf{I})$. We will say that f is *bounded when $x \to +\infty$* or *to the right* if and only if there exists an interval $\mathbf{I}' \subset \mathbf{I}$, unbounded to the right, a number $n \in \mathbb{N}$ and a function F, continuous and bounded on $\mathbf{I}'$, such that $f = \partial^n F$ on $\mathbf{I}'$. We denote by $\mathcal{B}_{+\infty}(\mathbf{I})$ the subset of $\mathcal{C}_\infty(\mathbf{I})$ comprising all distributions which are bounded when $x \to +\infty$, it being easy to see that $\mathcal{B}_{+\infty}(\mathbf{I})$ is a linear subspace of $\mathcal{C}_\infty(\mathbf{I})$. It is also clear that any distribution which is convergent when $x \to +\infty$ is bounded : $\Lambda_{+\infty}(\mathbf{I}) \subset \mathcal{B}_{+\infty}(\mathbf{I})$. On the other hand we have:

Proposition 2.40 *If f is a distribution bounded to the right, φ a multiplier at the point $+\infty$ and $\lim_{x \to +\infty} \varphi(x) = 0$, then $\lim_{x \to +\infty} \varphi(x)f(x) = 0$.*

Proof: The proof is similar to the proof of proposition 2.15: the property being clearly true when F is a function which is continuous and bounded to the right, suppose that it is also true for $f \in \mathcal{B}_{+\infty}(\mathbf{I})$ (with any multiplier converging to zero) and let $g = \partial f$. Then the equality

$$\varphi g = \varphi \partial f = \partial(\varphi f) - \psi f,$$

where $\psi = \partial \varphi - \varphi$, shows that the proposition also holds for g, and this ends the proof. $\quad\square$

Now let f be a distribution and φ a function of class $\mathcal{C}^\infty$, both defined on intervals unbounded to the right; we will say that f is *dominated by φ* (when $x \to +\infty$ or to the right) and will write $f \in \mathbf{O}(\varphi)$ if there exist an interval unbounded to the right $\mathbf{I}$ and a distribution $f_0 \in \mathcal{C}_\infty(\mathbf{I})$, bounded to the right and such that $f = \varphi f_0$ on $\mathbf{I}$. Hence, for example, we will have $f \in \mathbf{O}(1)$ if f is a distribution bounded when $x \to +\infty$.

Let φ be any function which is $\mathcal{C}^\infty$ on an interval unbounded to the right. If $f \in \mathbf{o}(\varphi)$ then we also have $f \in \mathbf{O}(\varphi)$, that is, $\mathbf{o}(\varphi) \subset \mathbf{O}(\varphi)$; if $\alpha, \beta \in \mathbb{C}$ and $f, g \in \mathbf{o}(\varphi)$ then $\alpha f + \beta g \in \mathbf{o}(\varphi)$ and similarly for $\mathbf{O}(\varphi)$.

On the other hand we have:

Proposition 2.41 *If $\alpha, \beta \in \mathbb{R}$, $\alpha < \beta$ and $f \in \mathbf{O}(x^\alpha)$, then $f \in \mathbf{o}(x^\beta)$.*

Proposition 2.42 *If $\alpha \in \mathbb{R}$ and $f \in \mathbf{O}(x^\alpha)$ then $\mathbf{D}f \in \mathbf{O}(x^{\alpha-1})$; similarly, from $f \in \mathbf{o}(x^\alpha)$ it follows that $\mathbf{D}f \in \mathbf{o}(x^{\alpha-1})$.*

Proposition 2.41 is an immediate consequence of proposition 2.40. For proposition 2.42 note that from $f = x^\alpha f_0$ it follows that

$$\mathbf{D}f = x^{\alpha-1}(\alpha f_0 + x\mathbf{D}f_0) = x^{\alpha-1}(\partial f_0 + (\alpha - 1)f_0).$$

When f_0 is a distribution bounded to the right, or convergent to 0 when $x \to +\infty$, the same is true for the distribution $\partial f_0 + (\alpha - 1)f_0$. It follows that $\mathbf{D}f \in \mathbf{O}(x^{\alpha-1})$ in the first case and $\mathbf{D}f \in \mathbf{o}(x^{\alpha-1})$ in the second case.

The concept of a distribution bounded at $-\infty$, or to the left or to the right of a point $a \in \mathbb{R}$, is defined similarly. For example, we will say that the distribution f, defined on the interval $\mathbf{I} =]a, b[$ is *bounded when $x \to a$* if and only if there exists $n \in \mathbb{N}$ and a function F, continuous on $\mathbf{I}$ and bounded on an interval $]a, c[$ for some $c \in]a, b[$, such that $f = \partial_a^n F$. In the more general case, when f is defined on an interval $\mathbf{J}$ for which a is a limit point but not the right-hand end-point, we will say that f is *bounded when $x \to a^+$* if the restriction of f to the interval $\mathbf{J} \cap]a, +\infty[$ is bounded when $x \to a$ in the sense of the previous definition.

If $f \in \mathcal{C}_\infty(\mathbb{R})$ is a bounded distribution when $x \to +\infty$ and when $x \to -\infty$, we will say simply that f is bounded when $x \to \infty$: $f \in \mathcal{B}_\infty(\mathbb{R})$.

We also define in an obvious way the notations $\mathbf{o}(\varphi)$ and $\mathbf{O}(\varphi)$ in the cases $x \to -\infty$, $x \to a^+$, etc., and without any difficulty we can verify that similar propositions to 2.40 and 2.42 hold, while 2.41, under the hypothesis that $a \in \mathbb{R}$, clearly assumes the form

Proposition 2.43 *If $\alpha, \beta \in \mathbb{R}$, $\beta < \alpha$ and $f \in \mathbf{O}(x^\alpha)$ then $f \in \mathbf{o}(x^\beta)$.*

We intend now to define the notion of integral of a distribution on an interval in several situations.

To start with let $\mathbf{I}$ be an interval of $\mathbb{R}$, a an interior point of $\mathbf{I}$ and f a distribution defined on $\mathbf{I}$ and precontinuous at the point a. Given an arbitrary complex number k, there exists one and only one distribution $g \in \mathcal{C}_\infty(\mathbf{I})$, satisfying the following conditions:

$$\mathbf{D}g = f, \quad g(a) = k. \tag{2.30}$$

(The uniqueness is obvious, given that two primitives of f differ necessarily by a constant; to see the existence take an arbitrary primitive h of f and set $g = h - h(a) + k$.)

In particular, there is a unique (first) primitive of f which is zero at the point a, and it is natural to denote this primitive by any one of the symbols

$$\int_a^{\hat{x}} f, \quad \text{or} \quad \int_a^{\hat{x}} f(t)\, dt$$

(in which we will generally omit the hat over the symbol x); hence the solution of the problem (2.30) will take the form:

$$g(x) = k + \int_a^x f(t)\, dt.$$

We do not generally assume that the distribution g is continuous at any point x other than a; however, if g is continuous at a certain point b, it will also be natural to write:

$$g(b) = k + \int_a^b f(t)\,dt,$$

and this corresponds to defining the integral of f on the interval bounded by a and b by the generalised Barrow's Formula

$$\int_a^b f(t)\,dt = g(b) - g(a).$$

Other situations may be considered similarly: for example, if f is a distribution pre-convergent when $x \to a^+$ (that is, if there exists a primitive of f which is convergent when $x \to a^+$, where a is a limit point of $\mathbf{I}$ but not the right-hand end-point of $\mathbf{I}$), we may denote by

$$\int_{a+}^x f(t)\,dt$$

the primitive of f which has limit zero at a^+. Further in the case where f is also preconvergent when $x \to b^-$, where we may have either $b \neq a$ or $b = a$, we set by definition:

$$\int_{a+}^{b^-} f(t)\,dt = g(b^-) - g(a^+),$$

where g is any primitive of f.

If $a < b$ we may refer to $\int_{a+}^{b^-} f$ as the integral of f over the interval $]a, b[$. Also, provided the limits concerned do exist, the integrals

$$\int_{a-}^{b^-} f(x)\,dx = g(b^-) - g(a^-),$$

$$\int_{a-}^{b^+} f(x)\,dx = g(b^+) - g(a^-),$$

will similarly be referred to as the integrals of f over the intervals $[a, b[$ and $[a, b]$ respectively. In any case we will clearly have:

$$\int_{a-}^{b^+} f(x)\,dx = -\int_{b+}^{a^-} f(x)\,dx, \qquad \int_{a+}^{b^+} f(x)\,dx = -\int_{b+}^{a^+} f(x)\,dx, \quad \text{etc.}$$

provided that one of the integrals mentioned in each equality does exist. Thus, we have, for example:

$$\int_{0+}^{0^-} \delta(x)\,dx = -1, \qquad \int_{0+}^{1} \delta(x)\,dx = 0,$$

and

$$\int_{0-}^{1} \delta(x)\,dx = 1, \qquad \int_{0-}^{1} \delta'(x)\,dx = 0;$$

however, the integrals

$$\int_0^1 \delta(x)\,dx \quad \text{and} \quad \int_0^1 \delta'(x)\,dx,$$

do not exist.

When there is no danger of confusion, the integral of f over an interval $\mathbf{I}$ may alternatively be denoted by symbols such as

$$\int_{\mathbf{I}} f, \quad \int_{\mathbf{I}} f(x)\,dx.$$

Clearly we may also consider unbounded intervals: for example, the integral of f over the interval $]a, +\infty[$ will exist if f is preconvergent at a^+ and at $+\infty$. Under this hypothesis, if we have $\mathbf{D}g = f$, then

$$\int_{a+}^{+\infty} f(x)\,dx \;=\; g(+\infty) - g(a^+).$$

If f is preconvergent both at $+\infty$ and $-\infty$ we will have (with $\mathbf{D}g = f$)

$$\int_{\mathbf{R}} f \;=\; \int_{-\infty}^{+\infty} f(x)\,dx \;=\; g(+\infty) - g(-\infty).$$

We have, for example

$$\int_{-\infty}^{+\infty} \delta(x)\,dx \;=\; 1$$

and, if $\alpha \in \mathbb{R}\backslash\{0\}$,

$$\int_{-\infty}^{+\infty} e^{i\alpha x}\,dx \;=\; 0. \tag{2.31}$$

In case $\alpha = 0$, the last integral does not exist even in the distributional setting because no primitives of the function $\mathbf{1}$ will have a limit at $+\infty$ or at $-\infty$. Thus suppose, for example, that the function x had a limit in the sense of distributions (as defined in section 2.2) when $x \to +\infty$. Then the function $\mathbf{1} = \frac{1}{x} \cdot x$ would tend to 0 when $x \to +\infty$, in accordance with proposition 2.15 and this is absurd.

We will say that a distribution f is *integrable* over an interval $\mathbf{I}$ if and only if the integral of f over $\mathbf{I}$ exists.

It is easy to see that if F is a function with indefinite integral on an interval $\mathbf{I}$, the distribution which is identified with F will be integrable over any compact interval contained in $\mathbf{I}$, its integral being equal to the usual integral of the function F. Moreover, if the integral of F over some non-compact interval contained in $\mathbf{I}$ exists in either the proper or improper sense of function theory[11] then the integral of the corresponding distribution also exists and has the same value.

One of the most immediate properties of the integral is its linearity: if f and g are two integrable distributions on an interval $\mathbf{I}$ and if $\alpha, \beta \in \mathbf{C}$ then $\alpha f + \beta g$ is a distribution which is integrable on $\mathbf{I}$ and

$$\int_{\mathbf{I}} (\alpha f + \beta g) \;=\; \alpha \int_{\mathbf{I}} f + \beta \int_{\mathbf{I}} g.$$

---

[11]It should be noted that this concept of integral also generalises the concept of Lebesgue integral and of the conditionally convergent improper integrals.

By appealing to the final part of the first example of section 1.7, it is easy to see that if a is a point interior to $\mathbf{I}$ and $\varphi \in \mathcal{C}^\infty(\mathbf{I})$ we have

$$\int_{\mathbf{I}} \varphi(x)\delta^{(n)}(x-a)\,dx = \sum_{k=0}^{n}(-1)^k\binom{n}{k}\varphi^{(k)}(a)\int_{\mathbf{I}}\delta^{(n-k)}(x-a)\,dx$$
$$= (-1)^n\varphi^{(n)}(a).$$

Taking as a model the case of an integral over an interval of the form $\mathbf{I} =]a, b[$ (with $-\infty \le a < b \le +\infty$) we now state another very simple property:

Let $\varphi \in \mathcal{C}^\infty(\mathbf{I})$, $f \in \mathcal{C}_\infty(\mathbf{I})$ and consider the equality

$$\mathbf{D}(\varphi f) = \varphi\mathbf{D}f + \varphi'f;$$

if, among the three distributions $\mathbf{D}(\varphi f)$, $\varphi\mathbf{D}f$ and $\varphi'f$ two of them are preconvergent at a^+ and also two of them are preconvergent at b^-, then the limits $(\varphi f)(a^+)$ and $(\varphi f)(b^-)$ exist, the distributions $\varphi\mathbf{D}f$ and $\varphi'f$ are integrable over $\mathbf{I}$ and, moreover, the formula of integration by parts

$$\int_{a^+}^{b^-}\varphi\mathbf{D}f = \left[(\varphi f)(b^-) - (\varphi f)(a^+)\right] - \int_{a^+}^{b^-}\varphi'f,$$

holds.

For example, if $\alpha \ne 0$ and $g(x) = -\frac{i}{\alpha}xe^{i\alpha x}$, we will have (taking into account formula (2.31)):

$$\int_{-\infty}^{+\infty}xe^{i\alpha x}\,dx = g(+\infty) - g(-\infty) + \frac{i}{\alpha}\int_{-\infty}^{+\infty}e^{i\alpha x}\,dx = 0.$$

More generally it can be shown by induction that, for $\alpha \in \mathbb{R}\backslash\{0\}$ and $n \in \mathbb{N}$, we have

$$\int_{-\infty}^{+\infty}x^n e^{i\alpha x}\,dx = 0.$$

So far as the change of variable in the integral is concerned, we will state just one of the possible theorems which could be established as a prototype.

Let $\mathbf{J} =]a, b[$ $(a, b \in \mathbb{R}, a < b)$, and $\varphi : \mathbf{J} \to \mathbb{R}$ be a function of class $\mathcal{C}^\infty$ such that

(1) $\lim_{t\to a^+}\varphi(t) = -\infty$, $\lim_{t\to b^-}\varphi(t) = +\infty$,
(2) $\varphi'(t) > 0$ *for each* $t \in \mathbf{J}$,
(3) $\frac{\varphi(t)}{(t-a)\varphi'(t)} \in \mathcal{M}_a(\mathbf{J})$, $\frac{\varphi(t)}{(b-t)\varphi'(t)} \in \mathcal{M}_b(\mathbf{J})$.

Under these conditions, if f is a distribution integrable over $\mathbb{R}$, then the distribution $\varphi'(f \circ \varphi)$ is integrable over $\mathbf{J}$ and, moreover,

$$\int_{-\infty}^{+\infty}f(x)\,dx = \int_{a^+}^{b^-}f(\varphi(t))\varphi'(t)\,dt.$$

This result follows immediately from the application of propositions of the type 2.19 to an arbitrary primitive of the distribution f.

As an example let us show that there does not exist in the sense of distributions the integral

$$\int_{-\infty}^{+\infty} \sin \arctan x \, dx.$$

In fact, if this integral were to exist then so also would the integral obtained under the change of variable $x = \tan t$ where $t \in \,] - \frac{\pi}{2}, +\frac{\pi}{2}[$. This is easily seen to satisfy the conditions given above, and the resulting integral after the change of variable is

$$\int_{-\frac{\pi}{2}}^{+\frac{\pi}{2}} \frac{\sin t}{\cos^2 t} \, dt.$$

It is readily seen that the function $\frac{1}{\cos t}$, which is a primitive of the original integrand on $\,] - \frac{\pi}{2}, +\frac{\pi}{2}[$, does not converge in the sense of distributions when $t \to +\frac{\pi}{2}^-$ or when $t \to -\frac{\pi}{2}^+$.

Also by simple change of variables we can easily obtain the following formulas:

$$\int_{\mathbf{R}} f(x) \, dx \;=\; \int_{\mathbf{R}} f(x - a) \, dx$$

and

$$\int_{\mathbf{R}} f(x) \, dx \;=\; |c|^{-1} \int_{\mathbf{R}} f(cx) \, dx$$

which are valid for any distribution f, integrable on $\mathbb{R}$, and any real numbers a and c with $c \neq 0$.

We will now establish some important integrability criteria; we will take as a model the case of integrals on $\mathbb{R}$, the adaptations needed for the other cases being obvious. First of all

Proposition 2.44 *Let $f \in \mathcal{C}_\infty(\mathbb{R})$ and suppose that there exist $a, b \in \mathbb{R}$ such that $f = 0$ on $\,] - \infty, a[$ and on $\,]b, +\infty[$; Then f is integrable on $\mathbb{R}$.*

Proof: Let g denote a primitive of f. Then the hypothesis implies that there exist constants α and β such that $g = \alpha$ on $\,] - \infty, a[$ and $g = \beta$ on $\,]b, +\infty[$. We therefore have

$$\int_{-\infty}^{+\infty} f(x) \, dx \;=\; g(+\infty) - g(-\infty) \;=\; \beta - \alpha$$

which proves the proposition. $\qquad\qquad\Box$

Proposition 2.45 *If f is integrable on $\mathbb{R}$ then $f(x) \in \mathbf{O}(x^{-1})$ when x tends to ∞.*

Proof: If g is a primitive of f, we will have $g(x) \in \mathbf{O}(1)$ when $x \to \infty$ (that is, when $x \to +\infty$ and when $x \to -\infty$) and therefore, by proposition 2.42, $f = \mathbf{D}g \in \mathbf{O}(x^{-1})$ when $x \to \infty$. $\qquad\qquad\Box$

Proposition 2.46 *If there exists $\alpha < -1$ such that $f(x) \in \mathbf{O}(x^\alpha)$ when $x \to \infty$ then f is integrable on $\mathbb{R}$.*

Proof: First note that the following formula, similar to (2.4), is easily established by induction:

$$x^\alpha \partial^n f(x) = \sum_{k=0}^{n} c_k \partial^k \left[x^\alpha f(x) \right],$$

with

$$c_k = (-1)^{n-k} \binom{n}{k} \alpha^{n-k},$$

$\alpha \in \mathbb{R}$ and $n \in \mathbb{N}$.

Suppose now that $f(x) = x^\alpha f_0(x)$, with $\alpha < -1$ and $f_0(x) = \partial^n F_0(x)$, where $F_0(x)$ is a continuous function bounded on an interval $\mathbf{I} =]a, +\infty[$, where $a > 0$. Under these conditions, as we will show, f is preconvergent at $+\infty$. (A similar argument for an interval of the form $]-\infty, b[$, where $b < 0$, would establish preconvergence at $-\infty$). In fact, we will have on $\mathbf{I}$:

$$
\begin{aligned}
f(x) &= x^\alpha \partial^n F_0(x) = \sum_{k=0}^{n} c_k \partial^k \left[x^\alpha F_0(x) \right] \\
&= (-1)^n \alpha^n x^\alpha F_0(x) + \sum_{k=0}^{n} c_k \partial^k \left[x^\alpha F_0(x) \right],
\end{aligned}
$$

with $c_k = (-1)^k \binom{n}{k} \alpha^{n-k}$, and therefore, denoting by g a primitive of f

$$g(x) = (-1)^n \alpha^n \int_a^x t^\alpha F_0(t)\, dt + \sum_{k=1}^{n} c_k x \partial^{k-1} \left[x^\alpha F_0(x) \right] + C \tag{2.32}$$

for given constant $C \in \mathbf{C}$. However, if $|F_0(t)| \leq M$ for $t \in \mathbf{I}$, we will also have

$$\left| \int_a^x t^\alpha F_0(t)\, dt \right| \leq M x^{\alpha+1} \quad (x \in \mathbf{I})$$

and it follows that , for $\alpha < -1$, the first term of the right-hand side of (2.32) will converge to 0 when $x \to +\infty$ (in the sense of the functions and therefore also in the sense of the distributions). For the second term, again taking account of the formula referred to at the beginning of this proof, we get

$$\sum_{k=1}^{n} c_k x \partial^{k-1} \left[x^\alpha F_0(x) \right] = \sum_{k=1}^{n} c_k \sum_{l=0}^{k-1} (-1)^{k-l-1} \binom{k-1}{l} \partial^l \left[x^{\alpha+1} F_0(x) \right],$$

and this allows us to see that this term also tends to 0 (in the sense of distributions) when $x \to +\infty$. The proof is thus complete. $\square$

Propositions 2.45 and 2.46 may take other forms when one or other of the extremities of the interval of integration is finite. For example, for an integral of the form

$$\int_{a^-}^{+\infty} f(x)\, dx,$$

the existence of the integral implies that $f(x) \in \mathbf{O}\left((a-x)^{-1} \right)$ when $x \to a^-$ and $f(x) \in \mathbf{O}(x^{-1})$ when $x \to +\infty$; on the other hand, if $f(x) \in \mathbf{O}\left((a-x)^\alpha \right)$ when

$x \to a^-$ and $f(x) \in \mathbf{O}(x^\beta)$ when $x \to +\infty$, with $\alpha > -1$ and $\beta < -1$, then the integral certainly exists.

We end this section with an almost obvious remark on the relation between integration and trivial extension. First of all let f be a distribution defined on the interval $\mathbf{I} =]a, +\infty[$ and preconvergent when $x \to a$; then there will exist the trivial extension $\check{f}$ of f to $\mathbb{R}$ and, if g is an arbitrary primitive of f, it is easy to see that the relation (similar to that of Theorem 2.28)

$$\mathbf{D}\check{g} \;=\; \check{f} + g(a^+)\delta_{(a)}$$

holds. Thus $h = \check{g} - g(a^+)\mathbf{H}_{(a)}$ is a primitive of $\check{f}$ with limit zero at $-\infty$, and so $\check{f}$ is integrable on $\mathbb{R}$ if and only if the limit of $\check{g}$ (or of g) exists when $x \to +\infty$; that is, if and only if f is integrable on $\mathbf{I}$. When that is the case we will have

$$\begin{aligned}
\int_{\mathbf{R}} \check{f} \;&=\; h(+\infty) - h(-\infty) \\
&=\; \check{g}(+\infty) - g(a^+) \;=\; g(+\infty) - g(a^+) \;=\; \int_{\mathbf{I}} f.
\end{aligned}$$

Suppose now finally that f is a distribution defined on the bounded interval $\mathbf{J} =]a, b[$. In this case it will be natural to say that f *is trivially extendable to* $\mathbb{R}$ if and only if there exists a distribution $\check{f}$, precontinuous at a and b, such that

    (1) $\check{f} = 0$ on $]-\infty, a[$ and on $]b, +\infty[$,

    (2) $\check{f} = f$ on $\mathbf{J}$.

It is left as an exercise to the reader to confirm that a necessary and sufficient condition for f to be trivially extendable is that it should be preconvergent at a and b; that is, that it should be integrable on $\mathbf{J}$. In this case the trivial extension of f is unique and we have

$$\int_{\mathbf{R}} \check{f} \;=\; \int_{\mathbf{J}} f.$$

Chapter 3
Convergence of Distributions. Periodic Distributions. Global Distributions.

3.1 Convergence of Sequences of Distributions.

First recall that if E is a complex linear space then a *norm on E* is a map $x \rightsquigarrow \|x\|$ from E into $\mathbb{R}$ such that the following conditions are satisfied, for any x, y in E and $\alpha \in \mathbb{C}$:

n1. $\|x\| \geq 0$,

n2. $\|x\| = 0$ if and only if x is the zero in E,

n3. $\|x + y\| \leq \|x\| + \|y\|$,

n4. $\|\alpha x\| = |\alpha| \, \|x\|$.

Given a norm on E, topological notions such as open set, closed set, convergent sequence, etc. can immediately be defined on E. For example, we say that a sequence $(x_n)_{n \in \mathbb{N}}$ of terms in E *converges* to $u \in E$ if and only if

$$\lim_{n \to \infty} \|x_n - u\| = 0.$$

Similarly we say that a set $U \subset E$ is *closed* if and only if whenever a sequence $(x_n)_{n \in \mathbb{N}}$ of terms in U converges in E to u, then $u \in U$; and so on.

A particularly important example of a normed linear space (that is, a linear space provided with a norm) is the space $\mathcal{C}(K)$ comprising all (complex-valued) continuous functions defined on a compact interval $K \subset \mathbb{R}$, with the norm defined for each function $F \in \mathcal{C}(K)$ by

$$\|F\| = \max_{t \in K} |F(t)|.$$

When a sequence $(F_n)_{n \in \mathbb{N}}$ converges to a function F in the sense of this norm (that is, when, for each $\delta > 0$ there exists $p \in \mathbb{N}$ such that whenever $n > p$ we have

$$|F_n(t) - F(t)| < \delta$$

for each $t \in K$) it is usual to say that the sequence $(F_n)_{n \in \mathbb{N}}$ *converges uniformly to F in K*.

Another important notion to recall is that of continuous operator: if X and Y are linear normed spaces (or, more generally, spaces in which some specific concept of convergence of sequences is defined) we will say that $\Phi : X \to Y$ is a *continuous map*[1] (or *a continuous operator*) if and only if, whenever $(x_n)_{n\in\mathbb{N}}$ is a sequence in X converging to the vector u (in X), the transformed sequence $\Phi(x_n)$ converges to $\Phi(u)$ (in Y).

For example, supposing again that K is a (nondegenerate) compact interval of $\mathbb{R}$, let $X = \mathcal{C}(K)$ and $Y = \mathcal{C}^1(K)$, the first of these spaces having the norm defined above and the second with the *norm induced* by the first (that is, the restriction to Y of the norm considered in X). Given an arbitrarily chosen point $c \in K$, for each function $F \in \mathcal{C}(K)$ and each $t \in K$, set

$$\mathcal{J}F(t) \; = \; \int_c^t F(\tau)\,d\tau.$$

The symbol $\mathcal{J}$ then denotes a map from $\mathcal{C}(K)$ into $\mathcal{C}^1(K)$, and it is easy to see that if $(F_n)_{n\in\mathbb{N}}$ is a sequence converging in $\mathcal{C}(K)$ to F, that is if $(F_n)_{n\in\mathbb{N}}$ converges uniformly to F on K, then $(\mathcal{J}F_n)_{n\in\mathbb{N}}$ also converges in $\mathcal{C}^1(K)$; in fact, from the relations

$$\begin{aligned}
|\mathcal{J}F_n(t) - \mathcal{J}F(t)| \; &= \; \left| \int_c^t (F_n(\tau) - F(\tau))\,d\tau \right| \\
&\leq \; \left| \int_c^t |F_n(\tau) - F(\tau)|\,d\tau \right| \\
&\leq \; \left| \int_c^t \|F_n - F\|\,d\tau \right| \; \leq \; (b-a)\|F_n - F\|
\end{aligned}$$

which are valid for every $t \in K$, it follows at once that,[2]

$$\|\mathcal{J}F_n - \mathcal{J}F\| \; = \; \max_{t\in K} |\mathcal{J}F_n(t) - \mathcal{J}F(t)| \; \leq \; (b-a)\|F_n - F\|,$$

and therefore, if

$$\lim_{n\to\infty} \|F_n - F\| \; = \; 0$$

we also certainly have

$$\lim_{n\to\infty} \|\mathcal{J}F_n - \mathcal{J}F\| \; = \; 0.$$

It follows therefore that $\mathcal{J} : \mathcal{C}(K) \to \mathcal{C}^1(K)$ is a continuous operator. It is worth noting that, with the norms chosen as above, the operator $\mathbf{D} : \mathcal{C}^1(K) \to \mathcal{C}(K)$ *is not* continuous. If, for example, c is a point of K, the sequence

$$F_n(t) \; = \; \frac{1}{n} \; \sin[n(t + \pi - c)]$$

---

[1]More properly we should say "sequentially continuous". This is equivalent to "continuous" in the case of normed spaces, but may not be so when the convergence of sequences is not defined by a norm. Similarly, the notion of closed set referred to above would be better expressed by saying "sequentially closed set" instead.

[2]When there is no danger of confusion the same symbol may be used to denote norms defined in different spaces.

does converge uniformly on K to the null function, since for every $n \in \mathbf{N}$ we have $\|F_n\| \leq \frac{1}{n}$, but the sequence of derivatives

$$\mathbf{D}F_n(t) = \cos[n(t + \pi - c)]$$

does not converges in K (it diverges, for example, at the point c).

The continuity of an operator is a property of considerable importance, both from the theoretical point of view and in all kind of applications. For this reason, where an operator of particular importance is concerned, it is often convenient to adopt definitions of convergence in the domain and in the range spaces which ensure the continuity of that operator.

For example, to guarantee continuity of the operator $\mathbf{D} : \mathcal{C}^1(K) \to \mathcal{C}(K)$ we may be led to adopt in $\mathcal{C}^1(K)$ the norm defined by

$$\|F\| = \max\{\max_{t \in K} |F(t)|, \max_{t \in K} |F'(t)|\} \quad (F \in \mathcal{C}^1(K))$$

instead of the norm induced on $\mathcal{C}^1(K)$ by the $\mathcal{C}(K)$ norm. It is easy to see that for a sequence $(F_n)_{n \in \mathbf{N}}$ to converge to F in the space $\mathcal{C}^1(K)$ with that norm it is necessary and sufficient that both sequences $(F_n)_{n \in \mathbf{N}}$ and $(F'_n)_{n \in \mathbf{N}}$ do converge uniformly in K, the first to F and the second to F'. Clearly, with this new concept of convergence in $\mathcal{C}^1(K)$ (and with the original sense of convergence in the space $\mathcal{C}(K)$) the operator $\mathbf{D} : \mathcal{C}^1(K) \to \mathcal{C}(K)$ will be continuous.

Similarly, if we consider in the space $\mathcal{C}^p(K)$, $(p > 1)$ the convergence determined by the norm

$$\|F\| = \max\{\max_{t \in K} |F(t)|, \max_{t \in K} |F'(t)|, \ldots, \max_{t \in K} |F^{(p)}(t)|\}$$

then we accept that a sequence $(F_n)_{n \in \mathbf{N}}$ converges in $\mathcal{C}^p(K)$ if and only if each one of the sequences $(F_n)_{n \in \mathbf{N}}, (F'_n)_{n \in \mathbf{N}}, \ldots, (F_n^{(p)})_{n \in \mathbf{N}}$ converges uniformly in K to $F, F', \ldots, F^{(p)}$ respectively. It is obvious that the operator $\mathbf{D}$ will be a continuous map from $\mathcal{C}^p(K)$ into $\mathcal{C}^{p-1}(K)$; more generally, $\mathbf{D}^j : \mathcal{C}^p(K) \to \mathcal{C}^{p-j}(K)$ will be a continuous map for $j = 0, 1, \ldots, p$.

Now consider the space $\mathcal{C}^\infty(K)$, comprising all functions which are infinitely smooth on K. It is worthwhile to ask in what sense a sequence $(F_n)_{n \in \mathbf{N}}$ may be said to converge to a limit F in this space. The most convenient answer is that $(F_n)_{n \in \mathbf{N}}$ converges to F in $\mathcal{C}^\infty$ when, for every $j \in \mathbf{N}$, the sequence $(\mathbf{D}^j F_n)_{n \in \mathbf{N}}$ converges uniformly to $F^{(j)}$ on K. It is then clear that with such definition for convergence, the operator $\mathbf{D}$ will be a continuous map from $\mathcal{C}^\infty(K)$ into itself.

It may be proved that there does not exist any norm over $\mathcal{C}^\infty(K)$ which determines the notion of convergence just referred to. Nevertheless it is this a concept of sequential convergence which it is convenient to adopt in $\mathcal{C}^\infty$: a sequence converges in $\mathcal{C}^\infty(K)$ if and only if it converges in each one of the spaces $\mathcal{C}^p(K)$ of which $\mathcal{C}^\infty(K)$ is the intersection. Hence, for example, if K is a compact interval contained in the interior of the interval of convergence of a power series, the sequence constituted by the partial sums of that series will converge in $\mathcal{C}^{\infty}(K)$ to the sum of the series.

We now turn to questions of convergence in spaces of distributions. To begin with, consider the space $\mathcal{C}_1(K)$ comprising all first order derivatives of functions continuous on K. To ensure the continuity of the operator $\mathbf{D} : \mathcal{C}(K) \to \mathcal{C}_1(K)$ we surely require that a sequence $(f_n)_{n\in\mathbb{N}}$ of elements of $\mathcal{C}_1(K)$ does converge to $f \in \mathcal{C}_1(K)$ whenever some sequence whose elements are primitives of the distributions f_n converges uniformly to a primitive of f. (In such a case it is clear that the primitives of the f_n and of f must be functions continuous on K).

We will adopt precisely this condition as the basis for the definition of convergence in $\mathcal{C}_1(K)$ and a similar condition for the spaces $\mathcal{C}_p(K)$, with $p > 1$. If $f \in \mathcal{C}_p(K)$ and $(f_n)_{n\in\mathbb{N}}$ is a sequence in $\mathcal{C}_p(K)$, we will say that $(f_n)_{n\in\mathbb{N}}$ converges to f in $\mathcal{C}_p(K)$ if and only if there exist functions F and F_n in $\mathcal{C}(K)$ such that we have $\mathbf{D}^p F_n = f_n$ and $\mathbf{D}^p F = f$ and $(F_n)_{n\in\mathbb{N}}$ converges to F uniformly in K.[3] On the other hand, since $\mathcal{C}_\infty(K)$ is the union of all spaces $\mathcal{C}_p(K)$, $(p \in \mathbb{N})$ it will be natural to say that a sequence $(f_n)_{n\in\mathbb{N}}$ converges to f in $\mathcal{C}_\infty(K)$ if and only if there is convergence in the above sense in some space $\mathcal{C}_p(K)$; that is, if and only if there exist $p \in \mathbb{N}$ and functions F and F_n such that $f = \mathbf{D}^p F$, $f_n = \mathbf{D}^p F_n$ and $(F_n)_{n\in\mathbb{N}}$ does converge to F uniformly in K.[4]

Finally if $(f_n)_{n\in\mathbb{N}}$ is a sequence of distributions defined on the interval $\mathbf{I}$ (not necessarily compact) we shall say that $(f_n)_{n\in\mathbb{N}}$ converges in $\mathcal{C}_\infty(\mathbf{I})$ to the distribution $f \in \mathcal{C}_\infty(\mathbf{I})$ if and only if, for any compact interval $K \subset \mathbf{I}$, the sequence comprising all the restrictions to K of the distributions f_n converges, in the space $\mathcal{C}_\infty(K)$ to the restriction of f to K.[5]

Thus, for example, each of the sequences $\delta(x - n)$, $\delta^{(n)}(x - n)$ converges to the null distribution in $\mathcal{C}_\infty(\mathbb{R})$. To see this it it is enough to note that, for any positive real number s, the restriction of each of the distributions $\delta(x - n)$ and $\delta^{(n)}(x - n)$ to the interval $[-s, +s]$ is null, provided $n > s$. However the sequence whose general term is

$$\sum_{j=0}^{n} \delta^{(j)}(x - j)$$

does not converge in $\mathcal{C}_\infty(\mathbb{R})$, as again it is easy to see.

The concept of convergence in the space $\mathcal{C}_\infty(\mathbf{I})$ just defined clearly satisfies the following properties:

> i) for $F, F_n \in \mathcal{C}(\mathbf{I})$ if the sequence $(F_n)_{n\in\mathbb{N}}$ converges uniformly to F in each compact interval contained in $\mathbf{I}$, then the sequence $(F_n)_{n\in\mathbb{N}}$ converges to F in $\mathcal{C}_\infty(\mathbf{I})$.

---

[3]It is easy to see that the concept of convergence thus defined in $\mathcal{C}_p(K)$ is the one which is generated by the norm $\| f \| = \inf\{\| F \| : f = \mathbf{D}^p F\}$ ($f \in \mathcal{C}_p(K), F \in \mathcal{C}(K)$), where $\| F \| = \max_{t\in K} |F(t)|$.

[4]It may be proved that the convergence thus defined in $\mathcal{C}_\infty(K)$ is not derived from any norm.

[5]Other modes of convergence in $\mathcal{C}_\infty(\mathbf{I})$ are possible; the one introduced here is the most suitable for the applications to be considered in the sequel.

ii) if the sequence of distributions $(f_n)_{n\in\mathbb{N}}$ converges to the distribution f in $\mathcal{C}_\infty(\mathbf{I})$, then so also does the sequence $(\mathbf{D}f_n)_{n\in\mathbb{N}}$ to $\mathbf{D}f$ in the same space (that is, the operator $\mathbf{D}:\mathcal{C}_\infty(\mathbf{I})\to\mathcal{C}_\infty(\mathbf{I})$ is continuous).

Before considering the uniqueness of the limit of a sequence of distributions it is convenient to establish the folllowing lemma:

Lemma 3.1 *Let $p\in\mathbb{N}$ and $(P_n(t))_{n\in\mathbb{N}}$ be a sequence of polynomials in t with complex coefficients and with degree $< p$; let also $t_1, t_2, \ldots, t_p$ be distinct points in $\mathbb{R}$. Then if all the numerical sequences $(P_n(t_1))_{n\in\mathbb{N}}, \ldots$ $(P_n(t_p))_{n\in\mathbb{N}}$ are convergent the sequence $(P_n(t))_{n\in\mathbb{N}}$ will also converge in $\mathbb{R}$ to a polynomial $P(t)$ of degree $< p$, the convergence being uniform in each compact interval of $\mathbb{R}$. Moreover, for each $j = 0, 1, \ldots, p-1$ the sequence constituted by the coefficients of t^j in the polynomials $P_n(t)$ will converge to the coefficient of t^j in the polynomial $P(t)$.*

Proof: Denote by a_n^j $(n\in\mathbb{N}, j = 0, 1, \ldots, p-1)$ the coefficient of the power t^j in the polynomial $P_n(t)$; for any n and any $l = 1, \ldots, p$ we will have:

$$a_n^0 + a_n^1 t_l + \cdots + a_n^{p-1} t_l^{p-1} \;=\; P_n(t_l). \tag{3.1}$$

Since $t_1, t_2, \ldots, t_p$ are distinct points in $\mathbb{R}$, the (Vandermonde) determinant

$$\begin{vmatrix} 1 & t_1 & \cdots & t_1^{p-1} \\ 1 & t_2 & \cdots & t_2^{p-1} \\ & & \cdots & \\ 1 & t_p & \cdots & t_p^{p-1} \end{vmatrix}$$

is not null and the system (3.1) allows us to express each one of the coefficients a_n^j as a linear combination of $P_n(t_1), \ldots, P_n(t_p)$ (with the coefficients independent of n). The existence of the limits $a^j = \lim_{n\to\infty} a_n^j$ then follows immediately. Setting

$$P(t) \;=\; a^0 + a^1 t + \cdots + a^{p-1} t^{p-1}$$

we will have for any t such that $|t| \le c$ (where c denotes an arbitrary positive real number):

$$|P_n(t) - P(t)| \;\le\; \sum_{j=0}^{p-1} |a_n^j - a^j| c^j.$$

Hence, given $\varepsilon > 0$ and defining

$$K \;=\; \max_j c^j$$

it will be enough to determine r, so that for $n > r$ we have $|a_n^j - a^j| < \varepsilon/pK$ for all $j = 0, 1, \ldots, p-1$, to obtain:

$$|P_n(t) - P(t)| \;\le\; \sum_{j=0}^{p-1} \frac{\varepsilon}{pK} K = \varepsilon.$$

The proof is then complete. $\qquad\square$

We now recall that from the continuity of the operator $\mathcal{J} : \mathcal{C}(K) \to \mathcal{C}^1(K)$ (where K is a compact interval) it follows that if the sequence $(F_n)_{n\in\mathbb{N}}$, with terms in $\mathcal{C}(K)$, converges uniformly on K to the function F then for $r \in \mathbb{N}$ the sequence $(\mathcal{J}^r F_n)_{n\in\mathbb{N}}$ also converges uniformly in K to $\mathcal{J}^r F$. From this it follows easily that if the sequences $(f_n)_{n\in\mathbb{N}}$ and $(g_n)_{n\in\mathbb{N}}$ converge to f and g, respectively, in $\mathcal{C}_\infty(K)$, we can deduce the existence of a natural number p (the same in both cases) and of functions $F, G, F_n, G_n \in \mathcal{C}(K)$ such that $f_n = \mathbf{D}^p F_n$, $f = \mathbf{D}^p F$, $g_n = \mathbf{D}^p G_n$, $g = \mathbf{D}^p G$ and the sequences $(F_n)_{n\in\mathbb{N}}$ and $(G_n)_{n\in\mathbb{N}}$ converge uniformly on K to F and G, respectively.

We may now state the following theorem which expresses the uniqueness of the limit:

Theorem 3.2 *Let $\mathbf{I}$ be an interval of $\mathbb{R}$, f and g two distributions defined on $\mathbf{I}$ and suppose that the sequence $(f_n)_{n\in\mathbb{N}}$ converges in $\mathcal{C}_\infty(\mathbf{I})$ to f and also to g. Then $f = g$.*

Proof: We will prove that in every compact interval $K \subset \mathbf{I}$ we have $f = g$, which is equivalent to the thesis of the theorem (cf. 1.11). From the hypothesis there follows the existence of a natural number p and functions $F, G, F_n, G_n \in \mathcal{C}(K)$ such that on K we have $f = \mathbf{D}^p F$, $g = \mathbf{D}^p G$ and $f_n = \mathbf{D}^p F_n = \mathbf{D}^p G_n$ with $(F_n)_{n\in\mathbb{N}}$ converging to F and $(G_n)_{n\in\mathbb{N}}$ to G uniformly on K. Hence the sequence $(F_n - G_n)_{n\in\mathbb{N}}$, whose terms are polynomials of degree $< p$ will converge to the function $F - G$ which, according to lemma 3.1, will be a polynomial of degree $< p$. Hence

$$f = \mathbf{D}^p F = \mathbf{D}^p G + \mathbf{D}^p(F - G) = \mathbf{D}^p G = g$$

as was to be proved.   □

Instead of saying that the sequence $(f_n)_{n\in\mathbb{N}}$ converges to the distribution f we will often say that f is the *limit* of f_n; more briefly still, we may sometimes write this as $f_n \to f$, or as $\lim f_n = f$.

We now establish several simple propositions which relate convergence to operations defined on spaces of distributions. Some of these will only be stated for particular situations since no greater generality will be required in the sequel.

Proposition 3.3 *If $\alpha, \beta \in \mathbf{C}$, $f_n \to f$ and $g_n \to g$ in $\mathcal{C}_\infty(\mathbf{I})$, then we have that $h_n = \alpha f_n + \beta g_n \to h = \alpha f + \beta g$.*

Proof: If $K \subset \mathbf{I}$ is a compact interval then there will exist $p \in \mathbb{N}$ and $F, G, F_n, G_n \in \mathcal{C}(K)$ such that (on K): $f = \mathbf{D}^p F$, $f_n = \mathbf{D}^p F_n$, $g = \mathbf{D}^p G$ and $g_n = \mathbf{D}^p G_n$ where $(F_n)_{n\in\mathbb{N}}$ converges to F and $(G_n)_{n\in\mathbb{N}}$ converges to G, uniformly on K. Under these conditions $H_n = \alpha F_n + \beta G_n$ will converge uniformly on K to $H = \alpha F + \beta G$. The result follows immediately since $h = \mathbf{D}^p H$ and $h_n = \mathbf{D}^p H_n$.   □

Proposition 3.4 *Let $(\varphi_n)_{n\in\mathbb{N}}$ be a sequence of elements in $\mathcal{C}^\infty(\mathbf{I})$ and φ a function in $\mathcal{C}^\infty(\mathbf{I})$ and suppose that, for each $j \in \mathbb{N}$ the sequence $(\varphi_n^{(j)})_{n\in\mathbb{N}}$ converges to $\varphi^{(j)}$ uniformly in each compact interval contained in $\mathbf{I}$; also let $(f_n)_{n\in\mathbb{N}}$ be a sequence of distributions defined on $\mathbf{I}$ converging to* *the distribution f (that is, $f = \lim f_n$). Then the sequence $(\varphi_n f_n)_{n \in \mathbb{N}}$ is convergent and $\varphi f = \lim_{n \to \infty}(\varphi_n f_n)$.*

Proof: If $K \subset \mathbf{I}$ is a compact interval there will exist $p \in \mathbb{N}$ and $F, F_n \in \mathcal{C}(K)$ such that (in K) $f = \mathbf{D}^p F$ and $f_n = \mathbf{D}^p F_n$ where $(F_n)_{n \in \mathbb{N}}$ converges to F uniformly on K. Taking into account the relations

$$\left| \varphi_n^{(j)}(t) F_n(t) - \varphi^{(j)}(t) F(t) \right| \leq$$

$$\left| \varphi_n^{(j)}(t) \right| \, |F_n(t) - F(t)| + \left| \varphi_n^{(j)}(t) - \varphi^{(j)}(t) \right| \, |F(t)|$$

and the fact that the convergence of $(F_n)_{n \in \mathbb{N}}$ to F and of $(\varphi_n^{(j)})_{n \in \mathbb{N}}$ to $\varphi^{(j)}$ is uniform on K (from which in particular it follows that for some M we have $|\varphi_n^{(j)}(t)| \leq M$ for all $n \in \mathbb{N}$ and any $t \in K$) it is easily shown that the sequence $(\varphi_n^{(j)} F_n)_{n \in \mathbb{N}}$ converges to $\varphi^{(j)} F$, uniformly on K (for each $j \in \mathbb{N}$).

Moreover, from the continuity of the operator $\mathbf{D} : \mathcal{C}_\infty(K) \to \mathcal{C}_\infty(\mathcal{K})$, the proposition 3.3 and the formula

$$\varphi_n f_n = \varphi_n \mathbf{D}^p F_n = \sum_{j=0}^{p} (-1)^j \tbinom{p}{j} \mathbf{D}^{p-j} \left(\varphi_n^{(j)} F_n \right),$$

it is easy to see that the sequence $(\varphi_n f_n)_{n \in \mathbb{N}}$ converges (in each compact interval $K \subset \mathbf{I}$) to the distribution

$$\sum_{j=0}^{p} (-1)^j \tbinom{p}{j} \mathbf{D}^{p-j} \left(\varphi^{(j)} F \right),$$

that is, to φf.

Therefore it follows that $(\varphi_n f_n)_{n \in \mathbb{N}}$ converges to φf in $\mathcal{C}_\infty(\mathbf{I})$. $\square$

Proposition 3.5 *Let $\mathbf{I}$ and $\mathbf{I}'$ be two intervals of $\mathbb{R}$, and let φ and φ_n ($n \in \mathbb{N}$) be functions of class $\mathcal{C}^\infty$ defined on $\mathbf{I}'$ with values in $\mathbf{I}$, whose first derivatives do not vanish at any point of $\mathbf{I}'$. Suppose also that, for each $j \in \mathbb{N}$, the sequence $(\varphi_n^{(j)})_{n \in \mathbb{N}}$ converges to $\varphi^{(j)}$ uniformly on the compacts contained in $\mathbf{I}'$. Finally let $(f_n)_{n \in \mathbb{N}}$ be a sequence of distributions converging in $\mathcal{C}_\infty(\mathbf{I})$ to the distribution f. Under these conditions the sequence $(f_n \circ \varphi_n)_{n \in \mathbb{N}}$ converges in $\mathcal{C}_\infty(\mathbf{I})$ to $f \circ \varphi$.*

Proof: We will suppose that $\varphi'(t) > 0$ for each $t \in \mathbf{I}'$ (the case where $\varphi'(t) < 0$ is similar). Under the conditions of the hypothesis, the functions φ_n' will also be positive from a certain index on. Neglecting, if necessary, some initial terms, we may suppose that $\varphi_n'(t) > 0$ for each $t \in \mathbf{I}'$ and each $n \in \mathbb{N}$.

Now let $K' = [\alpha, \beta]$ be an arbitrary compact interval contained in $\mathbf{I}'$ and set

$$a = \varphi(\alpha), \quad a_n = \varphi_n(\alpha), \quad b = \varphi(\beta), \quad b_n = \varphi_n(\beta),$$

$$K = [a, b] \quad \text{and} \quad K_n = [a_n, b_n].$$

Since $(a_n)_{n \in \mathbb{N}}$ is a sequence of points of $\mathbf{I}$ converging to a, the set formed by all the terms a_n together with the limit a is a compact contained in $\mathbf{I}$ and therefore has a minimum $a^* \in \mathbf{I}$;

similarly, the set formed by the numbers b_n and b has a maximum b^*. Hence $K^* = [a^*, b^*]$ will be a compact interval contained in $\mathbf{I}$ such that $K \subset K^*$ and $K_n \subset K^*$ for each n.

Now represent the distributions f and f_n in K^* in the form

$$f = \mathbf{D}_x^p F \quad \text{and} \quad f_n = \mathbf{D}_x^p F_n$$

with $F, F_n \in \mathcal{C}(K^*)$ and $(F_n)_{n \in \mathbf{N}}$ converging to F uniformly in K^*.

The relation

$$|F_n(\varphi_n(t)) - F(\varphi(t))| \leq$$
$$|F_n(\varphi_n(t)) - F(\varphi_n(t))| + |F(\varphi_n(t)) - F(\varphi(t))|$$

allow us to see easily that, under the conditions of the hypothesis, the sequence $(F_n \circ \varphi_n)_{n \in \mathbf{N}}$ converges to $F \circ \varphi$ uniformly in K'. Hence the continuity of the operator

$$\mathbf{D}_t : \mathcal{C}_\infty(K') \rightarrow \mathcal{C}_\infty(K')$$

the proposition 3.4 and the equality

$$f_n \circ \varphi_n = \left(\frac{1}{\varphi_n'} \, \mathbf{D}_t \right)^p (F_n \circ \varphi_n)$$

show immediately that the sequence $(f_n \circ \varphi_n)_{n \in \mathbf{N}}$ converges in $\mathcal{C}_\infty(K')$ to the distribution

$$\left(\frac{1}{\varphi'} \, \mathbf{D}_t \right)^p (F \circ \varphi),$$

that is, to the restriction to K' of the distribution $f \circ \varphi$. The proof is therefore complete. $\square$

Before giving some examples of sequences which converge in the space of distributions, it is convenient to introduce a very simple convergence criterion which is based on the notion of convergence in the mean. We start with the following definition:

Let $\mathbf{I}$ be an interval of $\mathbb{R}$, F and F_n $(n \in \mathbb{N})$ functions with indefinite integral in the interval $\mathbf{I}$ and K a compact interval contained in $\mathbf{I}$. We say that $(F_n)_{n \in \mathbf{N}}$ *converges in the mean* to the function F *in the interval* K if and only if the condition

$$\lim_{n \to \infty} \int_K |F_n(x) - F(x)| \, dx = 0$$

holds (in more intuitive terms: if and only if, over the interval K, the area lying between the graphs of the functions F_n and F tend to zero as $n \to \infty$).

It is easy to see that uniform convergence in K implies convergence in the mean in the same interval; the converse, however, is not true: for example, the sequence $(x^n)_{n \in \mathbf{N}}$ converges in the mean to the null function over the interval $[0, 1]$ but, obviously, the convergence is not uniform.[6]

We may now state the criterion referred to above.

---

[6]In fact it may be remarked that for convergence in the mean (as opposed to uniform convergence) there is no uniqueness of the limit: it may be proved that, if $(F_n)_{n \in \mathbf{N}}$ converges in the mean to F in K, then it also converges to any other function G which has the same indefinite integral over the interval K.

Proposition 3.6 *Let* **I** *be an interval of* $\mathbb{R}$, *and let* F *and* F_n *be functions with indefinite integral in* **I**. *If the sequence* $(F_n)_{n \in \mathbb{N}}$ *converges in the mean to* F *in each compact interval* $K \subset$ **I** *then it also converges in the space* $\mathcal{C}_\infty(\mathbf{I})$.

Proof: We will see that in each compact interval K contained in **I**, $(F_n)_{n \in \mathbb{N}}$ converges to F in $\mathcal{C}_1(K)$ and this implies the convergence of $(F_n)_{n \in \mathbb{N}}$ to F in $\mathcal{C}_\infty(\mathbf{I})$.

Let $c \in K$ and, for $x \in K$ and $n \in \mathbb{N}$, define:

$$\varphi(x) = \int_c^x F(t)\, dt, \qquad \varphi_n(x) = \int_c^x F_n(t)\, dt.$$

Then we will have $\varphi, \varphi_n \in \mathcal{C}(K)$ and, for any $x \in K$:

$$|\varphi_n(x) - \varphi(x)| \leq \left| \int_c^x |F_n(t) - F(t)|\, dt \right| \leq \int_K |F_n(t) - F(t)|\, dt$$

and so we also have

$$\|\varphi_n - \varphi\| = \max_{x \in K} |\varphi_n(x) - \varphi(x)| \leq \int_K |F_n(t) - F(t)|\, dt.$$

Since the integral tends to 0 by hypothesis, it follows that $(\varphi_n)_{n \in \mathbb{N}}$ converges to φ in $\mathcal{C}(K)$, and therefore, by the continuity of the operator **D**, $(F_n)_{n \in \mathbb{N}}$ converges to F in the space $\mathcal{C}_1(K)$. $\qquad\qquad\square$

If, for example, K is a compact interval of $\mathbb{R}$ we certainly have

$$\int_K \left| \mathbf{H}\left(x - \frac{1}{n} \right) - \mathbf{H}(x) \right| dx \; \leq \; \frac{1}{n}.$$

Hence, in $\mathcal{C}_\infty(\mathbb{R})$

$$\lim_{n \to \infty} \mathbf{H}\left(x - \frac{1}{n} \right) \; = \; \mathbf{H}(x)$$

and, taking into account the continuity of the operator **D**

$$\lim_{n \to \infty} \delta^{(p)}\left(x - \frac{1}{n} \right) \; = \; \delta^{(p)}(x)$$

for each $p \in \mathbb{N}$.

The application of the criterion 3.6 to the so-called Dirac sequences is of particular importance. A sequence $(K_n)_{n \in \mathbb{N}}$, where the terms are functions with indefinite integral in $\mathbb{R}$, is said to be a *Dirac sequence* if and only if it satisfies the following conditions:

d1) $K_n(x) \geq 0$, for each $x \in \mathbb{R}$;

d2) for any $n \in \mathbb{N}$ the improper integral

$$\int_{-\infty}^{+\infty} K_n(x)\, dx$$

converges, and its value is 1;

d3) for any positive numbers ε and ε', there exists p such that for $n > p$,

$$\int_{-\varepsilon}^{+\varepsilon} K_n(x)\,dx \;>\; 1 - \varepsilon'.$$

Intuitively, **d3)** imposes the condition that the area bounded by the graph of K_n and the x-axis (which according to **d2)**, is equal to 1 for every n) is, for all sufficiently large values of n, almost entirely concentrated in an arbitrarily narrow vertical strip which is symmetric with respect to the axis of ordinates.

We will now prove the following result:

> **Proposition 3.7** *Every Dirac sequence converges, in $\mathcal{C}_\infty(\mathbb{R})$, to the distribution δ.*

Proof: According to proposition 3.6 it will be enough to prove that, if $(K_n)_{n\in\mathbb{N}}$ is a Dirac sequence, there will exist a sequence formed by primitives of K_n which converges in the mean to the Heaviside function in any compact interval of $\mathbb{R}$.

Suppose then that we fix on some arbitrarily chosen compact K (which we may clearly assume to be of the form $[-s, +s]$ with $s > 0$), together with a number $\alpha > 0$. Setting

$$G_n(x) \;=\; \int_{-\infty}^{x} K_n(t)\,dt$$

and $\varepsilon = \min\{1, \frac{\alpha}{2s+2}\}$, determine p so that, for any $n > p$, we have

$$\int_{-\varepsilon}^{+\varepsilon} K_n(t)\,dt \;>\; 1 - \varepsilon.$$

We will have therefore $0 \le G_n(x) < \varepsilon$ for $x < -\varepsilon$ and $1 - \varepsilon < G_n(x) \le 1$ for $x > \varepsilon$, where G_n is an increasing function. Hence

$$\begin{aligned}
\int_{-s}^{+s} |G_n(x) - \mathbf{H}(x)|\,dx \;&=\; \int_{-s}^{-\varepsilon} G_n(x)\,dx + \int_{-\varepsilon}^{0} G_n(x)\,dx \\
&\quad + \int_{0}^{\varepsilon} (1 - G_n(x))\,dx + \int_{\varepsilon}^{s} (1 - G_n(x))\,dx \\
&\le\; \varepsilon(s - \varepsilon) + \varepsilon + \varepsilon + \varepsilon(s - \varepsilon) \;<\; 2s\varepsilon + 2\varepsilon \;\le\; \alpha,
\end{aligned}$$

and this proves that G_n converges in the mean to $\mathbf{H}$ in $[-s, +s]$. $\qquad\square$

As a first example of a Dirac sequence, consider the sequence whose nth term is the probability density of the normal distribution with mean value equal to zero and standard deviation σ_n:

$$K_n(x) \;=\; \frac{1}{\sigma_n\sqrt{2\pi}}\,e^{-\frac{x^2}{2\sigma_n^2}},$$

where $(\sigma_n)_{n\in\mathbb{N}}$ is a sequence of positive numbers converging to 0.

This sequence clearly satisfies **d1)**. Next, since the change of variable $x = -\sigma_n\sqrt{2}\,t$ transforms the integral of K_n over $\mathbb{R}$ into the product of $\frac{1}{\sqrt{\pi}}$ by the integral

$$\int_{-\infty}^{+\infty} e^{-t^2}\,dt$$

which has the value $\sqrt{\pi}$, the condition **d2)** is also satisfied (the probabilistic interpretation of the integral

$$\int_{-\infty}^{+\infty} K_n(x)\,dx$$

also leads to the same result).

Finally to check **d3)** it is enough to note that, using the change of variable referred to above and setting $\gamma_n = \frac{\varepsilon}{\sqrt{2}\sigma_n}$, we get

$$\int_{-\varepsilon}^{+\varepsilon} K_n(x)\,dx \;=\; 1 - 2\int_{+\varepsilon}^{+\infty} K_n(x)\,dx \;=\; 1 - \frac{2}{\sqrt{\pi}}\int_{\gamma_n}^{+\infty} e^{-t^2}\,dt$$

in which the last integral tends to zero when $n \to \infty$ due to the fact that $\sigma_n \to 0$ when $n \to \infty$.

It follows therefore that whenever $(\sigma_n)_{n\in N}$ is a sequence of positive numbers converging to zero, we have in the sense of distributions

$$\lim_{n\to\infty} \frac{1}{\sigma_n\sqrt{2\pi}}\, e^{-\frac{x^2}{2\sigma_n^2}} \;=\; \delta(x).$$

Other simple examples of Dirac sequences, as the reader may easily verify, are the sequences with general terms of the form

$$\frac{1}{\pi}\frac{n}{1+n^2x^2}, \qquad \frac{1}{\pi}\frac{n}{e^{nx}+e^{-nx}}$$

$$n\left[\mathbf{H}\left(x+\frac{1}{2n}\right) - \mathbf{H}\left(x-\frac{1}{2n}\right)\right], \qquad \frac{1}{2}\,n\,e^{-n|x|}.$$

All these sequences do converge in $\mathcal{C}_\infty(\mathbb{R})$ to the distribution δ. Since the terms of some of them are functions of class $\mathcal{C}^\infty$ it follows that (taking into account the continuity of the operator **D**) any derivative of the Dirac distribution is the limit, in $\mathcal{C}_\infty(\mathbb{R})$, of a sequence of infinitely smooth functions.

However it is possible to obtain a stronger result: specifically, it is possible to prove that any distribution f is the limit, in the space $\mathcal{C}_\infty(\mathbb{R})$, of a sequence of polynomials (a fact which is sometimes described by saying that the set of polynomials is *dense* in the space of distributions). We will give in this section the proof of this fact based on an important theorem of Weierstrass. For the proof of that theorem it is convenient now to introduce the following proposition:

Proposition 3.8 *Let $(K_n)_{n\in N}$ be a Dirac sequence and let F be a bounded function in $\mathcal{C}(\mathbb{R})$. Setting, for each $x \in \mathbb{R}$:*

$$F_n(x) \;=\; \int_{-\infty}^{+\infty} F(x-t)K_n(t)\,dt, \tag{3.2}$$

the sequence $(F_n)_{n\in N}$ will converge to F, uniformly in any compact of $\mathbb{R}$.

Proof: We start by noting that, for any $x \in \mathbb{R}$, the function $F(x - t)K_n(t)$ is integrable in t over each compact interval, since it is the product of two functions which satisfy that condition.[7] Moreover, the improper integral on the right-hand side of (3.2) converges since, if we take

$$M = \sup_{x \in \mathbf{R}} |F(x)|,$$

then the absolute value of the integrand will be bounded by $MK_n(t)$, a function with an absolutely convergent integral over $(-\infty, +\infty)$.

Since by one of the conditions of the definition of a Dirac sequence, we have

$$F(x) = \int_{-\infty}^{+\infty} F(x)K_n(t)\,dt \qquad (x \in \mathbb{R})$$

it follows that

$$
\begin{aligned}
|F_n(x) - F(x)| &= \left| \int_{-\infty}^{+\infty} [F(x - t) - F(x)]K_n(t)\,dt \right| \\
&\leq \int_{-\infty}^{+\infty} |F(x - t) - F(x)|K_n(t)\,dt.
\end{aligned}
\tag{3.3}
$$

Hence, given arbitrarily a compact $K \subset \mathbb{R}$ and a number $\varepsilon > 0$, it will be possible (since F is uniformly continuous on compacts) to find $\alpha > 0$ such that, for any t such that $|t| < \alpha$ and any $x \in K$, we have

$$|F(x - t) - F(x)| < \frac{\varepsilon}{2}.$$

On the other hand, recalling that

$$M = \sup_{x \in \mathbf{R}} |F(x)|$$

(and supposing that $M \neq 0$, otherwise the result would be trivial) it is possible to determine p such that, for $n > p$, we have

$$\int_{-\alpha}^{+\alpha} K_n(t)\,dt > 1 - \frac{\varepsilon}{4M}.$$

For $n > p$ and $x \in \mathbb{R}$, we will have therefore

$$
\int_{-\infty}^{-\alpha} |F(x - t) - F(x)|K_n(t)\,dt + \int_{+\alpha}^{+\infty} |F(x - t) - F(x)|K_n(t)\,dt
$$

$$
\leq 2M \left[1 - \int_{-\alpha}^{+\alpha} K_n(t)\,dt \right] < 2M\frac{\varepsilon}{4M} = \frac{\varepsilon}{2}
\tag{3.4}
$$

and, for $x \in K$

$$
\int_{-\alpha}^{+\alpha} |F(x - t) - F(x)|K_n(t)\,dt \leq \frac{\varepsilon}{2} \int_{-\alpha}^{+\alpha} K_n(t)\,dt \leq \frac{\varepsilon}{2}.
\tag{3.5}
$$

Thus it follows from (3.3), (3.4) and (3.5) that

$$|F_n(x) - F(x)| < \varepsilon$$

---

[7]The function F_n defined by the above formula is called the *convolution* of F and K_n (the defining formula is similar to (2.19), but in the present situation we do not work with continuous functions which are null to the left of the origin).

for every $x \in K$ and any $n > p$, and this proves that $(F_n)_{n \in \mathbb{N}}$ converges to F uniformly on K. $\qquad\square$

We may now state

Theorem 3.9 (Weierstrass Theorem.) *Let* $\mathbf{I}$ *be a compact interval* $[a, b]$ *of* $\mathbb{R}$. *If* F *is a continuous function on* $\mathbf{I}$, *then there exists a sequence of polynomials which converges to* F *uniformly on* $\mathbf{I}$.

Proof: Consider first the case when $a = 0$, $b = 1$ and $F(0) = F(1) = 0$. The extension to the general case is a simple exercise which will be left to the end of the proof.

Denote by $\check{F}$ the function defined on $\mathbb{R}$ given by

$$\check{F}(x) = \begin{cases} F(x) & \text{if} \quad x \in [0, 1] \\ 0 & \text{if} \quad x \notin [0, 1]. \end{cases}$$

Then $\check{F}$ is a bounded continuous function on $\mathbb{R}$. Next, for each positive integer number n, set

$$K_n(x) = \begin{cases} \frac{(1-x^2)^n}{c_n} & \text{if} \quad x \in [-1, +1] \\ 0 & \text{if} \quad x \notin [-1, +1], \end{cases}$$

where

$$c_n = \int_{-1}^{+1} (1 - x^2)^n \, dx.$$

We begin by showing that $(K_n)_{n \in \mathbb{N}}$ is a Dirac sequence. Clearly conditions **d1)** and **d2)** are satisfied; to prove **d3)**, since all the functions K_n are even and vanish outside the interval $[-1, +1]$, it is enough to show that, for any $\delta \in]0, 1[$,

$$\lim_{n \to \infty} \int_{\delta}^{1} K_n(x) \, dx = 0.$$

For this purpose, note that

$$\frac{c_n}{2} = \int_{0}^{1} (1 - x^2)^n dx = \int_{0}^{1} (1 - x)^n (1 + x)^n dx \geq \int_{0}^{1} (1 - x)^n dx = \frac{1}{n + 1}$$

and, therefore, $c_n \geq \frac{2}{n+1}$, for every n. It then follows that

$$\begin{aligned} \int_{\delta}^{1} K_n(x) \, dx &= \int_{\delta}^{1} \frac{(1 - x^2)^n}{c_n} \, dx \\ &\leq \int_{\delta}^{1} \frac{n + 1}{2} (1 - \delta^2)^n dx = \frac{n + 1}{2} (1 - \delta^2)^n (1 - \delta). \end{aligned}$$

Since $1 - \delta^2 \in]0, 1[$, the sequence $\frac{n+1}{2}(1 - \delta^2)^n$ tends to zero, and this allows us to see that condition **d3)** is also satisfied. Hence $(K_n)_{n \in \mathbb{N}}$ is a Dirac sequence.

Consider now the sequence of functions

$$\begin{aligned} P_n(x) &= \int_{-\infty}^{+\infty} \check{F}(x - t) K_n(t) \, dt \\ &= \int_{-\infty}^{+\infty} \check{F}(t) K_n(x - t) \, dt = \int_{0}^{1} F(t) K_n(x - t) \, dt \end{aligned}$$

which, according to 3.8, converges uniformly to $\check{F}$ on the compacts of $\mathbb{R}$ and, therefore, to F on $[0, 1]$. Taking into account the expression for K_n it is easy to see that $K_n(x - t)$ is a polynomial in x and t, representable in the form

$$K_n(x - t) = \pi_0(t) + \pi_1(t)x + \cdots + \pi_{2n}(t)x^{2n},$$

where $\pi_0, \pi_1, \ldots, \pi_{2n}$ are polynomials in t. It follows then that

$$P_n(x) = a_0 + a_1 x + \cdots + a_{2n} x^{2n},$$

with

$$a_j = \int_0^1 F(t)\pi_j(t)\, dt, \quad (j = 0, 1, \ldots, 2n).$$

Hence the sequence of polynomials $(P_n)_{n \in \mathbb{N}}$ converges to F, uniformly on $[0, 1]$.

Consider now the general case: $\mathbf{I} = [a, b]$, with $a < b$, F a continuous function on $\mathbf{I}$, but otherwise arbitrary, and set $t = \frac{x-a}{b-a}$. For $x \in [a, b]$, t will belong to the interval $[0, 1]$ and the function G defined on this interval by

$$G(t) = F[a + (b - a)t] - F(a) - [F(b) - F(a)]t \tag{3.6}$$

will be continuous and such that $G(0) = G(1) = 0$.

Thus there will exist a sequence of polynomials converging uniformly to G on $[0, 1]$ and therefore, for any arbitrarily chosen $\varepsilon > 0$, there will certainly exist a polynomial $Q(t)$ such that, for $t \in [0, 1]$

$$|Q(t) - G(t)| < \varepsilon.$$

We will therefore have, for any $x \in [a, b]$

$$\left| Q\left(\frac{x - a}{b - a}\right) - G\left(\frac{x - a}{b - a}\right) \right| < \varepsilon \tag{3.7}$$

and, since by (3.6),

$$G\left(\frac{x - a}{b - a}\right) = F(x) - F(a) - \frac{F(b) - F(a)}{b - a}(x - a) \tag{3.8}$$

then if we set

$$P(x) = F(a) + \frac{F(b) - F(a)}{b - a}(x - a) + Q\left(\frac{x - a}{b - a}\right) \tag{3.9}$$

$P(x)$ will be a polynomial. Further (3.8) and (3.9) together give, for any $x \in [a, b]$,

$$P(x) + G\left(\frac{x - a}{b - a}\right) = F(x) + Q\left(\frac{x - a}{b - a}\right).$$

From this equality, and using (3.7), it follows that $|F(x) - P(x)| < \varepsilon$ for every $x \in [a, b]$. Thus, given $\varepsilon > 0$, there exists a polynomial $P(x)$ such that

$$|F(x) - P(x)| < \varepsilon \quad (x \in [a, b]).$$

It follows that to obtain a sequence of polynomials $(P_n)_{n \in \mathbb{N}}$ which is uniformly convergent to F in the interval $[a, b]$ it will be enough to determine, for each positive integer n, a polynomial P_n satisfying the condition

$$|F(x) - P_n(x)| < \frac{1}{n}$$

for all $x \in [a,b]$. $\square$

Now let f be an arbitrary distribution defined on $\mathbb{R}$ and let $f = \mathbf{D}^p F$ with F in $\mathcal{C}(\mathbb{R})$. In accordance with the result just proved, for each positive integer n we may determine a polynomial $P_n(x)$ such that for $|x| \leq n$

$$|F(x) - P_n(x)| < \frac{1}{n}.$$

Clearly the sequence $(P_n)_{n \in \mathbb{N}}$ converges to the function F uniformly on compacts of $\mathbb{R}$ and, therefore, also in $\mathcal{C}_\infty(\mathbb{R})$. Taking into account the continuity of the operator $\mathbf{D}$ we obtain

$$\lim_{n \to \infty} \mathbf{D}^p P_n = \mathbf{D}^p F = f$$

where, clearly, the terms of the sequence $(\mathbf{D}^p P_n)_{n \in \mathbb{N}}$ are polynomials. We may therefore state:

> **Proposition 3.10** *Any distribution defined on $\mathbb{R}$ is the limit, in $\mathcal{C}_\infty$, of a sequence of polynomials.*

Note that if the distribution f is not itself a polynomial then the sequence $(P_n)_{n \in \mathbb{N}}$ must contain polynomials of arbitrarily large degree. This follows from the following proposition which will be useful in the sequel:

> **Proposition 3.11** *If the sequence $(P_n)_{n \in \mathbb{N}}$, whose terms are all polynomials of degree $\leq r$, converges in $\mathcal{C}_\infty(\mathbb{R})$ to the distribution f, then f is a polynomial of degree $\leq r$ and, moreover, for each $x \in \mathbb{R}$, we have*
>
> $$\lim_{n \to \infty} P_n(x) = f(x)$$
>
> *where the convergence is uniform on the compacts of $\mathbb{R}$.*

Proof: It is enough to prove that, in any compact interval $K \subset \mathbb{R}$, the sequence $(P_n)_{n \in \mathbb{N}}$ converges uniformly to a polynomial P: the identity principle for polynomials then will guarantee that P is independent of the interval K. The degree of P will be $\leq r$ as a consequence of 3.1 and, since $(P_n)_{n \in \mathbb{N}}$ converges uniformly on the compacts of $\mathbb{R}$ it will also converge in $\mathcal{C}_\infty(\mathbb{R})$, from which by (3.2) it follows that $f = P$.

Now let K be a compact interval of $\mathbb{R}$. There will exist $p \in \mathbb{N}$ and functions $F, F_n \in \mathcal{C}(K)$ such that F_n converges uniformly to F on K, $f = \mathbf{D}^p F$ and $P_n = \mathbf{D}^p F_n$.

If P_n is a polynomial of degree $\leq r$, F_n will be a polynomial of degree $\leq p + r$ and the same will be true for the limit F; it follows that $f = \mathbf{D}^p F$ will be a polynomial of degree $\leq r$.

On the other hand, as seen in 3.1, the convergence of $(F_n)_{n \in \mathbb{N}}$ to F implies, for each j, the convergence of the sequence of the coefficients of x^j in the polynomial F_n to the coefficient of x^j in F. From this it can be deduced immediately that the sequence $(F_n^{(p)})_{n \in \mathbb{N}} = (P_n)_{n \in \mathbb{N}}$ will converge to the polynomial $P = f$ uniformly in K. $\square$

It is convenient to detach the following particular case of the proposition 3.11:

Proposition 3.12 *If the sequence $(c_n)_{n\in\mathbb{N}}$, of constant distributions, converges in $\mathcal{C}_\infty(\mathbf{I})$ to the distribution f, then f is a constant c and the numerical sequence $(c_n)_{n\in\mathbb{N}}$ converges to c in the usual sense.*

To conclude we prove the following proposition:

Proposition 3.13 *Let $\mathbf{I}$ be an interval of $\mathbb{R}$, $(f_n)_{n\in\mathbb{N}}$ a sequence of distributions converging in $\mathcal{C}_\infty(\mathbf{I})$; then there exist distributions g_n such that $\mathbf{D}g_n = f_n$, where $(g_n)_{n\in\mathbb{N}}$ is a convergent sequence.*

Proof: The result is immediate when $\mathbf{I}$ is compact; having chosen $p \in \mathbb{N}$ ($p \geq 1$) and $F_n \in \mathcal{C}(\mathbf{I})$, with $(F_n)_{n\in\mathbb{N}}$ converging uniformly on $\mathbf{I}$ and $f_n = \mathbf{D}^p F_n$ it will be enough to set $g_n = \mathbf{D}^{p-1} F_n$.

When $\mathbf{I}$ is not compact, choose arbitrarily distributions h_n such that $\mathbf{D}h_n = f_n$ (on $\mathbf{I}$) and a compact (non-degenerate) interval $K \subset \mathbf{I}$. Since there are primitives, on K, of the distributions f_n forming a convergent sequence, there will necessarily exist constants c_n such that $(h_n + c_n)_{n\in\mathbb{N}}$ is a convergent sequence in $\mathcal{C}_\infty(K)$. We will see that, under these conditions, the sequence $(h_n + c_n)_{n\in\mathbb{N}}$ will also converge in the space $\mathcal{C}_\infty(\mathbf{I})$.

In fact, if K^* is an arbitrary compact interval contained in $\mathbf{I}$ (which we may suppose to be such that $K^* \supset K$) we may see similarly that there exist constants c_n^* for which the sequence $(h_n + c_n^*)_{n\in\mathbb{N}}$ converges in $\mathcal{C}_\infty(K^*)$; however, since $(h_n + c_n)_{n\in\mathbb{N}}$ and $(h_n + c_n^*)_{n\in\mathbb{N}}$ both converge in $\mathcal{C}_\infty(K)$ the same will be true for the sequence of the differences $(c_n - c_n^*)_{n\in\mathbb{N}}$. Hence, taking into account 3.12, there follows the convergence in the usual sense of the sequence of numbers $(c_n - c_n^*)_{n\in\mathbb{N}}$ and therefore also the convergence in $\mathcal{C}_\infty(K^*)$ of the sequence whose general term is $h_n + c_n = (h_n + c_n^*) + (c_n - c_n^*)$.   □

3.2 Series of Distributions. Convolution of Distributions with Support Lower or Upper Bounded. Applications to Difference Equations.

The definition of convergence of a series of distributions is now obvious: if $(f_n)_{n\in\mathbb{N}}$ is a sequence of distributions defined on an interval $\mathbf{I}$, we will say that the series

$$\sum_{n=1}^{\infty} f_n = f_1 + f_2 + \cdots + f_n + \cdots\cdots$$

is *convergent* (in $\mathcal{C}_\infty(\mathbf{I})$) if and only if the sequence of partial sums $g_n = f_1 + f_2 + \cdots + f_n$ is convergent. Under this hypothesis, the limit of $(g_n)_{n\in\mathbb{N}}$ is, by definition, the *sum of the series*.

As an immediate consequence of the linearity and continuity of the derivation operator, we have:

Proposition 3.14 *If the series $\sum f_n$ converges and has g as its sum, then the series $\sum \mathbf{D}f_n$ also converges and its sum is $\mathbf{D}g$. That is,*

$$\mathbf{D}\sum f_n = \sum \mathbf{D}f_n,$$

whenever the series $\sum f_n$ is convergent.

Also it is clear that for the series $\sum f_n$ to converge to the sum g it is necessary and sufficient that, for any compact interval $K \subset \mathbf{I}$, there exist a number $p \in \mathbb{N}$ and functions $G, F_n \in \mathcal{C}(K)$ such that $g = \mathbf{D}^p G$ and $f_n = \mathbf{D}^p F_n$, where the series $\sum F_n$ converges uniformly on K to the sum function G.

It is easy to see that if $(a_n)_{n \in \mathbb{N}}$ is an arbitrary sequence of complex numbers and T a positive real number, then the series

$$\sum_{n=1}^{\infty} a_n \delta(x - nT)$$

will always be convergent in $\mathcal{C}_\infty(\mathbb{R})$, and its sum will be the derivative of the function defined on $\mathbb{R}$ by the series

$$\sum_{n=1}^{\infty} a_n \mathbf{H}(x - nT)$$

(a function which is obviously integrable over any compact interval and therefore identifiable with an element in $\mathcal{C}_\infty(\mathbb{R})$).

In many situations it is important to consider series indexed by the set of integers, $\mathbb{Z}$. Naturally, we say that a series $\sum_{n \in \mathbb{Z}} f_n$ converges if and only if both series

$$f_0 + f_1 + f_2 + \cdots + f_n + \cdots\cdots$$
$$f_{-1} + f_{-2} + f_{-3} + \cdots + f_{-n} + \cdots\cdots$$

are convergent. Under this hypothesis, the sum of the series is, by definition, the sum of the two last series. And clearly the commutativity of the symbols $\mathbf{D}$ and $\sum$ remains in this case.

As an example, consider the series

$$\sum_{n \in \mathbb{Z}} c_n e^{in\omega x},$$

with $\omega > 0$. It is easy to prove that a sufficient condition for the convergence of this series is that its coefficients c_n constitute a *tempered sequence* (indexed by $\mathbb{Z}$), that is, that there exist a positive real number M and a natural number p so that, for every $n \neq 0$ we have:

$$|c_n| \leq M|n|^p.$$

In fact, if this condition is satisfied, the series

$$\sum_{n \in \mathbb{Z} \setminus \{0\}} \frac{c_n}{(in\omega)^{p+2}} e^{in\omega x} \tag{3.10}$$

will converge uniformly on $\mathbb{R}$ since we then have

$$\left| \frac{c_n}{(in\omega)^{p+2}} e^{in\omega x} \right| \leq \frac{M}{\omega^{p+2}} \frac{1}{n^2}$$

and the series $\sum \frac{1}{n^2}$ is convergent.

The sum of the series in (3.10) will therefore be a function continuous on $\mathbb{R}$ and, denoting it by G, we will have:

$$\sum_{n \in \mathbb{Z}} c_n e^{in\omega x} = c_0 + \sum_{n \in \mathbb{Z} \setminus \{0\}} \frac{c_n}{(in\omega)^{p+2}} \mathbf{D}^{p+2} e^{in\omega x} = c_0 + \mathbf{D}^{p+2} G(x).$$

Now let us return to the algebra $\mathcal{C}_\infty^+(\mathbb{R})$, introduced in section 2.4 and recall that, if $f, g \in \mathcal{C}_\infty^+(\mathbb{R})$ with $f = \mathbf{D}^p F$, $g = \mathbf{D}^q G$ and $F, G \in \mathcal{C}^+(\mathbb{R})$, we had set by definition:

$$f * g = \mathbf{D}^{p+q}(F * G)$$

where

$$F * G(x) = \int_0^x F(x - t)G(t)\, dt = \int_0^x F(t)G(x - t)\, dt, \quad x \in \mathbb{R}.$$

Suppose now that we have $f = 0$, not only on $]-\infty, 0[$, but in an interval of the form $]-\infty, c]$ where c is a positive number. In such a case, from the equality $f = \mathbf{D}^p F$, it follows that the restriction of F to $]-\infty, c]$ must be a polynomial; since $F(x) = 0$ for $x \leq 0$ then we also have $F(x) = 0$ for all $x \leq c$ and therefore we have

$$F * G(x) = \int_0^x F(t)G(x - t)\, dt = 0$$

for all $x \leq c$. If we let ρ denote (temporarily) the restriction operator for the interval $]-\infty, c]$, then we have

$$\rho(f * g) = \rho\left[\mathbf{D}^{p+q}(F * G)\right] = \mathbf{D}^{p+q}[\rho(F * G)] = 0.$$

Hence, the convolution $f * g$ of two distributions null to the left of the origin is null on the interval $]-\infty, c]$ whenever one of the two factors is null on $]-\infty, c]$. As an immediate consequence we have

Proposition 3.15 *Let $f_1, f_2, g \in \mathcal{C}_\infty^+(\mathbb{R})$ and $c > 0$; if $f_1 = f_2$ on $]-\infty, c]$, then we also have $f_1 * g = f_2 * g$ on $]-\infty, c]$.*

Each element g of the algebra $\mathcal{C}_\infty^+(\mathbb{R})$ defines a map Φ_g from $\mathcal{C}_\infty^+$ into itself according to:

$$\Phi_g(f) = f * g.$$

We intend to prove now that, if we consider in $\mathcal{C}_\infty^+(\mathbb{R})$ the notion of convergence of sequences induced by the convergence in $\mathcal{C}_\infty(\mathbb{R})$ then the map Φ_g is continuous. More precisely, we will prove that, whenever $(f_n)_{n \in \mathbb{N}}$ is a sequence with terms in $\mathcal{C}_\infty^+(\mathbb{R})$ converging (in $\mathcal{C}_\infty(\mathbb{R})$) to the distribution f, then it will follow that $f \in \mathcal{C}_\infty^+(\mathbb{R})$ and that the sequence $(f_n * g)_{n \in \mathbb{N}}$ converges to $f * g$.

The fact that we have $f \in \mathcal{C}_\infty^+(\mathbb{R})$ whenever $f = \lim f_n$ and $f_n \in \mathcal{C}_\infty^+$ is an immediate consequence of the commutativity of the limit operator and the restriction operator for the interval $]-\infty, 0[$; this follows from proposition (3.5). To prove the convergence of $(f_n * g)_{n \in \mathbb{N}}$ to $f * g$, it is necessary first to establish the following two preliminary lemmas:

Lemma 3.16 *Let $F, G \in \mathcal{C}^+(\mathbb{R})$, and let $(F_n)_{n \in \mathbb{N}}$ be a sequence of functions in $\mathcal{C}^+(\mathbb{R})$ such that, for each $c > 0$, the sequence $(F_n)_{n \in \mathbb{N}}$ converges to F uniformly on $]-\infty, c]$. Then, the sequence $(F_n * G)_{n \in \mathbb{N}}$ also converges to $F * G$ uniformly on $]-\infty, c]$ (for each $c > 0$).*

Proof: Given $c > 0$ and $\varepsilon > 0$ set $N = \max_{x \leq c} |G(x)|$ and (using the fact that the convergence of F_n to F is uniform in $]-\infty, c]$) determine p so that, for any $n > p$ and any $x \leq c$, we have

$$|F_n(x) - F(x)| \; < \; \frac{\varepsilon}{Nc}.$$

Then we also have, for $x \leq c$ and $n > p$:

$$|F_n * G(x) - F * G(x)| \;\; = \;\; \left| \int_0^x [F_n(x - t) - F(x - t)]\, G(t)\, dt \right|$$

$$\leq \;\; c\,\frac{\varepsilon}{Nc}\, N \;\; = \;\; \varepsilon,$$

and the proof is therefore complete. $\square$

In the lemma that follows we will prove a result which is subject to an extra condition to be relaxed at a later stage.

Lemma 3.17 *Let g, f and (for all $n \in \mathbb{N}$) f_n be distributions in $\mathcal{C}_\infty^+$, and suppose that there exists $c \geq 0$ such that, for each $n \in \mathbb{N}$, we have $f_n = 0$ on $]c, +\infty[$. Then, if $\lim_{n \to \infty} f_n = f$, it follows that $\lim_{n \to \infty}(f_n * g) = f * g$.*

Proof: Let d be any real number greater than c and set $K = [-d, +d]$; from the convergence of $(f_n)_{n \in \mathbb{N}}$ to f in $\mathcal{C}_\infty(\mathbb{R})$ it follows that there exists $p \in \mathbb{N}$ and functions $\Phi_n, \Phi \in C(K)$ such that $f_n = \mathbf{D}^p \Phi_n$, $f = \mathbf{D}^p \Phi$ and the sequence $(\Phi_n)_{n \in \mathbb{N}}$ converges uniformly in K to Φ.

Since $f_n = 0$ (and therefore also $f = 0$) in $[-d, 0[$ there will exist polynomials P_n, P (defined on $\mathbb{R}$ and of degree $< p$) such that $\Phi_n = P_n$ and $\Phi = P$ on $[-d, 0]$; similarly, since $f_n = f = 0$ on $]c, d]$, there will also exist polynomials Q_n, Q of degree $< p$, such that $\Phi_n = Q_n$ and $\Phi = Q$ on $[c, d]$. Since $(P_n)_{n \in \mathbb{N}}$ converges to P uniformly on $[-d, 0]$ and $(Q_n)_{n \in \mathbb{N}}$ converges to Q uniformly on $[c, d]$ then $(P_n)_{n \in \mathbb{N}}$ and $(Q_n)_{n \in \mathbb{N}}$ will converge to P and Q respectively, and the convergence will be uniform on compacts of $\mathbb{R}$.

Setting

$$F_n(x) \;\; = \;\; \begin{cases} 0 & \text{if } \; x \leq -d \\ \Phi_n(x) - P_n(x) & \text{if } \; |x| < d \\ Q_n(x) - P_n(x) & \text{if } \; x \geq d \end{cases}$$

$$F(x) \;\; = \;\; \begin{cases} 0 & \text{if } \; x \leq -d \\ \Phi(x) - P(x) & \text{if } \; |x| < d \\ Q(x) - P(x) & \text{if } \; x \geq d \end{cases}$$

we will have that $F_n, F \in \mathcal{C}^+(\mathbb{R})$, $f_n = \mathbf{D}^p F_n$ and $f = \mathbf{D}^p F$ where $(F_n)_{n \in \mathbb{N}}$ converges to F uniformly on $]-\infty, b]$, for each $b > 0$. Hence, if we have $g = \mathbf{D}^q G$, with $G \in \mathcal{C}^+(\mathbb{R})$, it follows from the foregoing lemma that $(F_n * G)_{n \in \mathbb{N}}$ will converge to $F * G$ uniformly in each interval of $\mathbb{R}$ bounded to the right and, therefore, also in $\mathcal{C}_\infty(\mathbb{R})$. From the continuity of the operator $\mathbf{D}$ it therefore follows that

$$\lim_{n\to\infty}(f_n * g) = \lim_{n\to\infty}[\mathbf{D}^{p+q}(F_n * G)]$$
$$= \mathbf{D}^{p+q}\lim_{n\to\infty}(F_n * G) = \mathbf{D}^{p+q}(F * G) = f * g$$

which proves the lemma. $\square$

We may now state

Theorem 3.18 *Let g, f and (for all $n \in \mathbb{N}$) f_n be distributions in $\mathcal{C}_\infty^+(\mathbb{R})$. If $(f_n)_{n\in\mathbb{N}}$ converges to f in $\mathcal{C}_\infty(\mathbb{R})$, then $(f_n * g)_{n\in\mathbb{N}}$ also converges in $\mathcal{C}_\infty(\mathbb{R})$ to $f * g$.*

Proof: It is enough to prove that for any $c > 0$, the sequence of the restrictions of $f_n * g$ to $]-\infty, c]$ converges to the restriction of $f * g$ to the same interval. Given some fixed c, let φ denote a function that satisfies the following conditions:[8]

(1) $\varphi \in C^\infty(\mathbb{R})$;

(2) $\varphi(x) = 1$ for $x \leq c$;

(3) there exists $d \in \mathbb{R}$ such that $\varphi(x) = 0$ for $x \geq d$.

By proposition 3.4 we have $\lim_{n\to\infty}(\varphi f_n) = \varphi f$ and therefore, by the lemma 3.17, $\lim_{n\to\infty}[(\varphi f_n) * g] = (\varphi f) * g$. Restricting each one of the distributions in the equality

$$f_n * g = (\varphi f_n) * g + [(1 - \varphi)f_n] * g$$

to the interval $]-\infty, c]$ it can be seen that the restriction of $(\varphi f_n) * g$ converges to the restriction of $(\varphi f) * g$ (that is, by proposition 3.15, to the restriction of $f * g$). Moreover, the restriction of $[(1 - \varphi)f_n] * g$ is null. It follows that the restriction of $f_n * g$ to $]-\infty, c]$ converges to the restriction of the distribution $f * g$, that is,

$$\lim_{n\to\infty}(f_n * g) = f * g$$

as was to be proved. $\square$

---

[8]To obtain an example of such a function φ, consider the function

$$\theta(x) = \begin{cases} A\exp\left(-\frac{1}{1-x^2}\right) & \text{if } |x| < 1 \\ 0 & \text{if } |x| \geq 1, \end{cases}$$

where A is a constant such that

$$\int_{-1}^{+1}\theta(x)\,dx = 1,$$

and set successively

$$\psi(x) = \int_{-\infty}^{x}\theta(t)\,dt \quad \text{and} \quad \varphi(x) = 1 - \psi(x - c - 1).$$

Since, for example

$$\lim_{n \to \infty} \delta(x - n) \; = \; 0$$

we will have, for any distribution $f \in \mathcal{C}_\infty^+$

$$\lim_{n \to \infty} [f * \delta(x - n)] \; = \; 0.$$

An important consequence of this theorem is the following: Let $\sum_{n=1}^{\infty} f_n$ be a convergent series of distributions null to the left of the origin, and let g be another distribution in the space $\mathcal{C}_\infty^+$. We then have:

$$g * \sum_{n=1}^{\infty} f_n \; = \; g * \lim_{n \to \infty} \left(\sum_{i=1}^{n} f_i \right) \; = \; \lim_{n \to \infty} \left(g * \sum_{i=1}^{n} f_i \right)$$

$$= \; \lim_{n \to \infty} \left(\sum_{i=1}^{n} g * f_i \right) \; = \; \sum_{n=1}^{\infty} (g * f_n).$$

We will now show that it is possible to enlarge the algebra $\mathcal{C}_\infty^+$ to get a new convolution algebra, which is more useful than $\mathcal{C}_\infty^+$ in several situations. This is the algebra of *distributions with support bounded below*.[9]

Let F and G be continuous functions, defined on $\mathbb{R}$, which are null for $x < 0$ and let h be a *positive* real number. Then $\tau_h F(\hat{x}) = F(\hat{x} - h)$ will also be an element in $\mathcal{C}_\infty^+(\mathbb{R})$ and moreover

$$\begin{aligned}
[(\tau_h F) * G](x) \; &= \; \int_0^x F(x - h - t)G(t)\, dt \\
&= \; \int_0^{x-h} F(x - h - t)G(t)\, dt + \int_{x-h}^x F(x - h - t)G(t)\, dt \\
&= \; [\tau_h(F * G)](x) + \int_{x-h}^x F(x - h - t)G(t)\, dt.
\end{aligned}$$

Since $F(x - h - t) = 0$ for $t > x - h$, the last integral is zero and so we may conclude that

$$(\tau_h F) * G \; = \; \tau_h(F * G).$$

For any two distributions $f, g \in \mathcal{C}_\infty^+$ ($f = \mathbf{D}^p F$, $g = \mathbf{D}^q G$, with $F, G \in \mathcal{C}^+(\mathbb{R})$) we will always have, for $h > 0$,

$$\begin{aligned}
(\tau_h f) * g \; &= \; \mathbf{D}^p(\tau_h F) * \mathbf{D}^q G \; = \; \mathbf{D}^{p+q}[(\tau_h F) * G] \\
&= \; \mathbf{D}^{p+q}[\tau_h(F * G)] \; = \; \tau_h \mathbf{D}^{p+q}(F * G) \; = \; \tau_h(f * g). \qquad (3.11)
\end{aligned}$$

Hence (as in the case of derivation) to apply the translation operator τ_h (with $h > 0$) to a convolution it is enough to apply the same translation to one of the factors; for example, with $f \in \mathcal{C}_\infty^+$, we have

$$\delta(\hat{x} - h) * f(\hat{x}) \; = \; \tau_h \delta * f \; = \; \tau_h(\delta * f) \; = \; \tau_h f \; = \; f(\hat{x} - h).$$

<hr>

[9]As will be seen in section 3.4, we define *support* of a distribution $f \in \mathcal{C}_\infty(\mathbb{R})$ to be the complement (with respect to $\mathbb{R}$) of the union of all open intervals on which the distribution f is null. Hence to say that f has support bounded below is the same as saying that $f = 0$ on some interval which is not bounded below.

For $h < 0$ the property (3.11) does not work because, in general, $\tau_h f$ is not null to the left of the origin even if f is. It is therefore natural to extend $\mathcal{C}_\infty^+$ so as to include those distributions which are obtained by applying operations of translation to elements of this space.

By convention we say that a distribution $f \in \mathcal{C}_\infty(\mathbb{R})$ has *support bounded below* if there exists $h \in \mathbb{R}$ such that $\tau_h f \in \mathcal{C}_\infty^+$ (or, equivalently, if there exists $b \in \mathbb{R}$ such that $f = 0$ on $]-\infty, b[$). By convention we also denote by $\mathcal{S}^+(\mathbb{R})$, (or simply by $\mathcal{S}^+$), the subspace of $\mathcal{C}_\infty(\mathbb{R})$ comprising all distributions with support bounded below. Clearly $\mathcal{C}_\infty^+$ is a linear subspace of $\mathcal{S}^+$.

We will now see that convolution, which is defined as a binary operation in $\mathcal{C}^+$, may be extended to $\mathcal{S}^+$. One way to do this which certainly seems to be the natural one, is the following: if $f, g \in \mathcal{S}^+$ with $f = \tau_h f_0$, $g = \tau_k g_0$, where h and k are real numbers and $f_0, g_0 \in \mathcal{C}_\infty^+$, then to define the convolution in $\mathcal{S}^+$ we set (using the same symbol $*$):

$$f * g = \tau_{h+k}(f_0 * g_0),$$

where $f_0 * g_0$ denotes the convolution in $\mathcal{C}_\infty^+$.

Since any distribution $f \in \mathcal{S}^+$ is representable in an infinity of ways in the form $f = \tau_h f_0$, with $f_0 \in \mathcal{C}_\infty^+$ and $h \in \mathbb{R}$, it is necessary to prove that the above definition is consistent.

Suppose, then, that we also have $f = \tau_l f_1$, with $f_1 \in \mathcal{C}_\infty^+$ and, for example, $l > h$. We will then have

$$\tau_h \tau_{l-h} f_1 = \tau_l f_1 = f = \tau_h f_0$$

and therefore, applying τ_{-h},

$$f_0 = \tau_{l-h} f_1.$$

It then follows that

$$\begin{aligned}
\tau_{l+k}(f_1 * g_0) &= \tau_{h+k}\tau_{l-h}(f_1 * g_0) \\
&= \tau_{h+k}[(\tau_{l-h} f_1) * g_0] = \tau_{h+k}(f_0 * g_0),
\end{aligned}$$

and this relation (together with the commutativity of the convolution in $\mathcal{C}_\infty^+$) allows us to conclude that the above definition is in fact consistent.

One can easily see that if $f, g \in \mathcal{S}^+$ then $f * g \in \mathcal{S}^+$, and also that the linear space $\mathcal{S}^+$, equipped with the convolution operation, is a commutative algebra. It is also immediate that, if $f, g \in \mathcal{S}^+$ and $h \in \mathbb{R}$, we have

$$\begin{aligned}
(\mathbf{D}f) * g &= \mathbf{D}(f * g) = f * (\mathbf{D}g) \\
(\tau_h f) * g &= \tau_h(f * g) = f * (\tau_h g).
\end{aligned}$$

Moreover, if $f = \tau_h f_0$, with $f_0 \in \mathcal{C}_\infty^+$, we will have

$$f * \delta = (\tau_h f_0) * \delta = \tau_h(f_0 * \delta) = \tau_h f_0 = f.$$

Hence, δ is the unit element for the convolution algebra $\mathcal{S}^+$. For the convolution with $\delta(\hat{x} - h)$, where h is now any point in $\mathbb{R}$, we have:

$$\delta(\hat{x} - h) * f(\hat{x}) \;=\; \tau_h(\delta * f) \;=\; f(\hat{x} - h).$$

More generally, for any $p \in \mathbb{N}$ and any $h \in \mathbb{R}$, we will have for any $f \in \mathcal{S}^+$:

$$f(\hat{x}) * \delta^{(p)}(\hat{x} - h) \;=\; \mathbf{D}^p f(\hat{x} - h).$$

We observe that, if $(f_n)_{n \in \mathbb{N}}$ is a sequence of distributions with terms in $\mathcal{S}^+$ which converges (in $\mathcal{C}_\infty$) to a distribution f, it may happen that f does not belong to $\mathcal{S}^+$: it is the case, for example, with the partial sums of the series $\sum_{n \in \mathbb{N}} \delta(x + n)$; however, if there exists a number $\alpha \in \mathbb{R}$ such that, for any n, we have $f_n = 0$ in $]-\infty, \alpha[$, then we also have $f = 0$ in $]-\infty, \alpha[$ and therefore $f \in \mathcal{S}^+$.

From the continuity of the translation operator (established by proposition 3.5) and the theorem 3.18, it is easy to see that if the condition referred to above is satisfied and $g \in \mathcal{S}^+$, then

$$\lim_{n \to \infty} (f_n * g) \;=\; f * g.$$

This result may not be true (even if we have $f \in \mathcal{S}^+$) if there does not exist $\alpha \in \mathbb{R}$ such that, for every $n \in \mathbb{N}$, $f_n = 0$ in $]-\infty, \alpha[$; for example, we have

$$\delta(x + n) * \mathbf{H}(x) \;=\; \mathbf{H}(x + n)$$

while

$$\lim_{n \to \infty} \delta(x + n) \;=\; \mathbf{0}, \quad \lim_{n \to \infty} \mathbf{H}(x + n) \;=\; \mathbf{1}.$$

Needless to say there is an entirely similar definition of a convolution product in the set $\mathcal{S}^-$ of distributions with support bounded above. This set comprises distributions of the form $\tau_h f$, where f is a distribution which is null to the right of the origin ($f \in \mathcal{C}_\infty^-$) and $h \in \mathbb{R}$. For $f, g \in \mathcal{S}^-$, $f = \tau_h f_0$, $g = \tau_k g_0$ with $f_0, g_0 \in \mathcal{C}_\infty^-$, we set naturally:

$$f * g \;=\; \tau_{h+k}\left(f_0 * g_0\right),$$

where $f_0 * g_0$ is the convolution in $\mathcal{C}_\infty^-$ defined in section 2.4. With this convolution the space $\mathcal{S}^-$ is a commutative algebra, in which δ is the unit element.

If $g, f_n \in \mathcal{S}^-$, and $\lim f_n = f$, and if there exists a number β such that, for each n, we have $f_n = 0$ in $]\beta, +\infty[$, then we still have $f \in \mathcal{S}^-$ and

$$\lim_{n \to \infty} (f_n * g) \;=\; f * g.$$

As a typical application, consider now the linear difference equation with constant coefficients:

$$a_0 u(x + n\tau) + a_1 u(x + n\tau - \tau) + \cdots$$

$$+ a_{n-1} u(x + \tau) + a_n u(x) \;=\; f(x), \tag{3.12}$$

where τ is a positive real number and $a_0, a_1, \ldots, a_n$ belong to $\mathbb{C}$, with $a_0 \neq 0$. Suppose in particular that f is a given distribution in the space $\mathcal{S}^+$ and u is a distribution to be determined also in $\mathcal{S}^+$. As will be seen shortly this equation always has a unique solution.

Equation (3.12) may be written in the form of a convolution equation in the algebra $\mathcal{S}^+$:

$$(a_0\delta(x + n\tau) + a_1\delta(x + n\tau - \tau) + \cdots + a_n\delta(x)) * u(x) = f(x). \tag{3.13}$$

Now, according to the remark made after the proof of proposition 2.36, to prove the existence of a unique solution in $\mathcal{S}^+$ it will be enough to show that the distribution

$$a(x) = a_0\delta(x + n\tau) + a_1\delta(x + n\tau - \tau) + \cdots + a_n\delta(x) \tag{3.14}$$

has an inverse in this algebra. In the first place it is easy to see that, denoting by $c_n, c_{n+1}, \ldots$ the complex numbers determined successively, from the given numbers $a_0, a_1, \ldots, a_n$, by the equations

$$\begin{aligned}
a_0 c_n &= 1 \\
a_0 c_{n+1} + a_1 c_n &= 0 \\
a_0 c_{n+2} + a_1 c_{n+1} + a_2 c_n &= 0 \\
&\cdots \\
a_0 c_{2n} + a_1 c_{2n-1} + \cdots + a_n c_n &= 0
\end{aligned}$$

and, for $p > 2n$,

$$a_0 c_p + a_1 c_{p-1} + \cdots + a_n c_{p-n} = 0,$$

then we will have

$$\sum_{i=0}^{n} a_i\delta(x + n\tau - i\tau) * \sum_{j=n}^{\infty} c_j\delta(x - j\tau) = \delta(x).$$

Then, once the constants c_j have been determined from the preceeding system, the inverse of the distribution $a(x)$ in the algebra $\mathcal{S}^+$ will be

$$a^{-1}(x) = \sum_{j=n}^{\infty} c_j\delta(x - j\tau),$$

and

$$u(x) = a^{-1}(x) * f(x)$$

will be the unique solution in $\mathcal{S}^+$ of the equation (3.12).

Suppose now that the right-hand side, f, of equation (3.13) belongs to the algebra $\mathcal{S}^-$ and that a solution u of the equation in the same algebra is required. Then the problem is to invert in $\mathcal{S}^-$ the distribution

$$a(x) = a_0\delta(x + n\tau) + a_1\delta(x + n\tau - \tau) + \cdots + a_n\delta(x).$$

Introducing the additional hypothesis $a_n \neq 0$, it is then easy to see that if the constants $c_0, c_{-1}, c_{-2}, \ldots$ are determined successively by the equations

$$
\begin{aligned}
a_n c_0 &= 1 \\
a_n c_{-1} + a_{n-1} c_0 &= 0 \\
a_n c_{-2} + a_{n-1} c_{-1} + a_{n-2} c_0 &= 0 \\
&\cdots \\
a_n c_{-n} + a_{n-1} c_{-n+1} + \cdots + a_0 c_0 &= 0
\end{aligned}
$$

and, for $p > n$,

$$
a_n c_{-p} + a_{n-1} c_{-p+1} + \cdots + a_0 c_{-p+n} = 0,
$$

then the distribution

$$
\sum_{j=0}^{\infty} c_{-j} \delta(x + j\tau)
$$

will be the inverse of $a(x)$ in the algebra $\mathcal{S}^-$.

Hence, assuming that the conditions $a_0 \neq 0$ and $a_n \neq 0$ are satisfied, the convolution equation (3.13) will have a unique solution either in the algebra $\mathcal{S}^+$ or in the algebra $\mathcal{S}^-$.

The possibility of being able to solve equation (3.12) in both algebras $\mathcal{S}^+$ and $\mathcal{S}^-$ leads naturally to the study of other problems with respect to the same equation in other spaces. One of the problems that can be stated in this setting is the following "*initial value*" problem:

Exercise 3.19 *To determine a solution* $u \in \mathcal{C}_\infty(\mathbb{R})$ *of the equation*

$$
a_0 u(x + n\tau) + a_1 u(x + n\tau - \tau) + \cdots + a_n u(x) = f(x),
$$

where $a_0, a_1, \ldots, a_n \in \mathbf{C}$, *with* $a_0 \neq 0$ *and* $a_n \neq 0$, *and where we now suppose that* f *is a given distribution in* $\mathcal{C}_\infty(\mathbb{R})$, *such that the following conditions hold:*

i) *for* $j = 0, 1, \ldots, n - 1$ *we must have*

$$
u = u_j \quad \text{in} \quad]j\tau, (j+1)\tau[
$$

where the distributions $u_j \in \mathcal{C}_\infty(]j\tau, (j+1)\tau[)$ *are given;*

ii) *the solution* u *must be a distribution precontinuous at the points* $j\tau$ *($j = 0, 1, \ldots, n$).*

It is clear that since $u(x)$ must be precontinuous at the points $0, \tau, 2\tau, \ldots, n\tau$, the distribution $u(x + \tau)$ must be precontinuous at the points $-\tau, 0, \tau, \ldots, (n-1)\tau$, etc.; finally $u(x + n\tau)$ must be precontinuous at the points $-n\tau, \ldots, -\tau$ and 0. Hence, if the problem has any solution, each term of the left-hand side of (3.12) will be a distribution precontinuous at the origin. It follows that a necessary condition for the existence of a solution is that f is a distribution precontinuous at the point 0; and also that (if there is a solution) we may cut off both members of (3.12) to obtain:

$$a_0 u(x + n\tau)\mathbf{H}(x) + a_1 u(x + n\tau - \tau)\mathbf{H}(x)$$

$$+ \cdots + a_n u(x)\mathbf{H}(x) = f(x)\mathbf{H}(x), \tag{3.15}$$

and

$$a_0 u(x + n\tau)\mathbf{H}^*(x) + a_1 u(x + n\tau - \tau)\mathbf{H}^*(x)$$

$$+ \cdots + a_n u(x)\mathbf{H}^*(x) = f(x)\mathbf{H}^*(x). \tag{3.16}$$

What is more, from the conditions **i)** and **ii)** of (3.19) it follows at once that each of the distributions u_j must be preconvergent at the points $j\tau$ and $(j + 1)\tau$. This is also a necessary condition for the existence of a solution to the problem.

Assuming this condition to be satisfied we will denote by v_j, for $j = 0, 1, \ldots, n - 1$, the common extension[10] of the distributions $u_0, u_1, \ldots, u_j$ to the interval $]0, (j + 1)\tau[$ which is precontinuous at the points $\tau, 2\tau, \ldots, j\tau$ (with $v_0 = u_0$) and by $\breve{w}_j$ the trivial extension to $\mathbb{R}$ of the distribution $w_j(x) = v_j(x + j\tau + \tau)$.

If u is a solution of the exercise 3.19 we must have (for $j = 1, \ldots, n$):

$$u(x + j\tau)\mathbf{H}(x) = u(x + j\tau)\mathbf{H}(x + j\tau) - \breve{w}_{j-1}, \tag{3.17}$$

$$u(x + j\tau)\mathbf{H}^*(x) = u(x + j\tau)\mathbf{H}^*(x + j\tau) - \breve{w}_{j-1}. \tag{3.18}$$

(To see this it is enough to note that, in any of these equalities, the distributions that appear in both members are identified in each of the intervals $] - \infty, -j\tau[,] - j\tau, 0[$, $]0, +\infty[$ and are precontinuous at the points $-j\tau$ and 0). From (3.15) and (3.17) it therefore follows that:

$$a_0 u(x + n\tau)\mathbf{H}(x + n\tau) + \cdots + a_n u(x)\mathbf{H}(x)$$

$$= f(x)\mathbf{H}(x) + (a_0\breve{w}_{n-1}(x) + a_1\breve{w}_{n-2}(x) + \cdots + a_{n-1}\breve{w}_0(x));$$

similarly, from (3.16) and (3.18) it follows that

$$a_0 u(x + n\tau)\mathbf{H}^*(x + n\tau) + \cdots + a_n u(x)\mathbf{H}^*(x)$$

$$= f(x)\mathbf{H}^*(x) - (a_0\breve{w}_{n-1}(x) + a_1\breve{w}_{n-2}(x) + \cdots + a_{n-1}\breve{w}_0(x));$$

It may therefore be concluded that, if there is a solution u to the problem (3.19), its truncation $u\mathbf{H}$ will be a solution of the equation in w

$$a_0 w(x + n\tau) + a_1 w(x + n\tau - \tau) + \cdots + a_n w(x)$$

$$= f(x)\mathbf{H}(x) + (a_0\breve{w}_{n-1}(x) + a_1\breve{w}_{n-2}(x) + \cdots + a_{n-1}\breve{w}_0(x)) \tag{3.19}$$

---

[10]It is easy to see that such an extension exists and is unique.

and its truncation $u\mathbf{H}^*$ will be a solution of the equation

$$a_0 w^*(x+n\tau) + a_1 w^*(x+n\tau-\tau) + \cdots + a_n w^*(x)$$

$$= f(x)\mathbf{H}^*(x) - (a_0\breve{w}_{n-1}(x) + a_1\breve{w}_{n-2}(x) + \cdots + a_{n-1}\breve{w}_0(x)). \qquad (3.20)$$

As seen above, the equation (3.19), equivalent to a convolution equation in the algebra $\mathcal{S}^+$, has a unique solution in this algebra; similarly, (3.20) has solution unique in $\mathcal{S}^-$. Then (if a solution u exists) the truncations $u\mathbf{H}$ and $u\mathbf{H}^*$ will be uniquely determined and therefore it follows that the solution $u = u\mathbf{H} + u\mathbf{H}^*$ will be unique.

We will prove now that if the necessary conditions referred to above are satisfied, that is, if $f \in \mathcal{C}_\infty(\mathbb{R})$ is a distribution precontinuous at the origin, and if u_j (for each $j = 0, 1, \ldots, n-1$) is a distribution defined on $]j\tau, (j+1)\tau[$ and preconvergent at each of the end-points of this interval, the problem (3.19) will in fact have a solution $u = w + w^*$, where we denote by w and w^* the solutions of (3.19) and (3.20) in $\mathcal{S}^+$ and $\mathcal{S}^-$, respectively.

First it is immediately obvious from the addition of (3.19) and (3.20) that:

$$a_0 u(x+n\tau) + a_1 u(x+n\tau-\tau) + \cdots + a_n u(x) = f(x),$$

which shows that $u = w + w^*$ is, in fact, a solution of the equation (3.12). Now to see that the initial conditions are satisfied set $\mathbf{I}_j =](-n+j)\tau, (-n+j+1)\tau[$, for $j = 0, 1, 2, \ldots, n-1$. On the interval $\mathbf{I}_0$ all terms of (3.19) are zero, with the possible exception of the terms $a_0 w(x+n\tau)$ and $a_0\breve{w}_{n-1}(x)$; hence, by restricting to $\mathbf{I}_0$ we get:

$$a_0 w(x+n\tau) = a_0\breve{w}_{n-1}(x) = a_0 w_{n-1}(x) = a_0 u_0(x+n\tau),$$

from which it follows that $w(x) = u_0(x)$ on $]0, \tau[$.

Restricting to $\mathbf{I}_1$ gives

$$a_0 w(x+n\tau) + a_1 w(x+n\tau-\tau) = a_0\breve{w}_{n-1}(x) + a_1\breve{w}_{n-2}(x),$$

and therefore

$$a_0 w(x+n\tau) + a_1 w(x+n\tau-\tau) = a_0 u_1(x+n\tau) + a_1 u_0(x+n\tau-\tau).$$

Since in $\mathbf{I}_1$ we have $w(x+n\tau-\tau) = u_0(x+n\tau-\tau)$, it follows that we will have $w(x+n\tau) = u_1(x+n\tau)$, that is, $w(x) = u_1(x)$ in $]\tau, 2\tau[$. In general, once the equalities

$$w(x) = u_j(x) \quad \text{on} \quad]j\tau, (j+1)\tau[$$

have been established for $j = 0, 1, \ldots, k-1$, (with $k < n$), the restriction to $\mathbf{I}_k$ gives

$$a_0 w(x+n\tau) + \cdots + a_k w(x+n\tau-k\tau) = a_0\breve{w}_{n-1}(x) + \cdots + a_k\breve{w}_{n-k-1}(x),$$

from which, taking into account the equalities already obtained, it is immediately deduced that $u = u_k$ on $]k\tau, (k+1)\tau[$.

Since the right-hand side of the equation (3.20) is a distribution null on $]0, +\infty[$ and since the unique solution w^* of this equation in the algebra $\mathcal{S}^-$ is the result of the convolution of that right-hand side with a distribution of the form $\sum_{j=0}^{\infty} c_{-j} \delta(x + j\tau)$, it may readily be concluded that we have $w^* = 0$ on $]0, +\infty[$ and therefore that the distribution $u = w + w^*$ in fact satisfies condition **i)** of (3.19).

To prove that condition **ii)** is also satisfied, consider again equality (3.19):

$$a_0 w(x + n\tau) + a_1 w(x + n\tau - \tau) + \cdots + a_n w(x)$$
$$= f(x)\mathbf{H}(x) + \left(a_0 \breve{w}_{n-1}(x) + a_1 \breve{w}_{n-2}(x) + \cdots + a_{n-1} \breve{w}_0(x)\right).$$

At the point $-n\tau$ all distributions appearing on either side of equation (3.19), with the possible exception of $a_0 w(x + n\tau)$ and $a_0 \breve{w}_{n-1}(x)$, are precontinuous (actually continuous and taking the value zero); since this last distribution is precontinuous at $-n\tau$, it follows that so also is $w(x + n\tau)$, that is that $w(x)$ is precontinuous at the point 0.

Now the distribution $a_1 w(x + n\tau - \tau)$ will be precontinuous at the point $-n\tau + \tau$ and, since the same is true for all the other terms in (3.19) with only the possible exception of $a_0 w(x + n\tau)$, this distribution also will be precontinuous at that point, and therefore w will be precontinuous at the point τ.

By repeating this reasoning successively with $x = -n\tau + 2\tau, \ldots, x = 0$, it follows without difficulty that w is precontinuous at the points $0, \tau, 2\tau, \ldots, n\tau$.

With respect to w^*, its precontinuity is obvious at the points $\tau, 2\tau, \ldots, n\tau$ where that distribution has the value zero. Further, taking into account this fact it is easily deduced from (3.20) that w^* is also precontinuous at the origin.

We may therefore conclude that the distribution $u = w + w^*$ is precontinuous at the points $0, \tau, 2\tau, \ldots, n\tau$ and this completes the proof of the following theorem:

Theorem 3.20 *For the exercise 3.19 to have a solution it is necessary that f should be precontinuous at the origin and that u_j should be preconvergent at the points $j\tau$ and $(j + 1)\tau$, for $j = 0, 1, \ldots, n - 1$. If these conditions are satisfied then the problem has one and only one solution.*

3.3   Periodic Distributions. Convolution. Fourier Series.

We will start this section by concentrating on a new aspect of the operation of convolution. This will be considered first in the context of continuous and periodic functions and then in that of periodic distributions. To begin with the development will follow the same lines as that of section 2.4, and this will allow us on occasion to shorten, or even dispense with, some of the proofs.

Given a positive real number T, we will denote by Π^T the (complex) linear space comprising all (complex-valued) periodic continuous functions on $\mathbb{R}$ which admit the period T. Simple examples of functions which belong to the space Π^T are the constant functions and the functions of the form $\cos(n\omega x)$, $\sin(n\omega x)$ and $\exp(in\omega x)$, with $\omega = \frac{2\pi}{T}$ and $n \in \mathbb{Z}$.

If, as we will suppose throughout the whole of this section, the period T is fixed once and for all, we may simply write Π instead of Π^T and abbreviate the expression *"function which admits the period T"* to *"periodic function"*.

For each pair (F, G) comprising two continuous periodic functions (with period T) and each $x \in \mathbb{R}$, we will set

$$F \otimes G(x) \ = \ \int_0^T F(x - t)G(t) \, dt. \tag{3.21}$$

By a simple inspection of this formula we may see that $F \otimes G : \mathbb{R} \to \mathbb{C}$ is a periodic function; it is also easy to see that it is continuous and therefore that it belongs to Π. In fact, given $\varepsilon > 0$, let $N = \max_{x \in \mathbb{R}} |G(x)|$. If $N = 0$ the result is trivial; otherwise choose $\gamma > 0$ such that, for $|h| < \gamma$ and $x \in \mathbb{R}$, we have

$$|F(x + h) - F(x)| \ < \ \frac{\varepsilon}{NT}$$

(which is always possible, since F is uniformly continuous on $\mathbb{R}$). We will then have, for any real x and for $|h| < \gamma$,

$$|F \otimes G(x + h) - F \otimes G(x)| \ = \ \left| \int_0^T [F(x - t + h) - F(x - t)]G(t) \, dt \right|$$

$$\leq \ \frac{\varepsilon}{NT} \, NT \ = \ \varepsilon.$$

Hence $(F, G) \rightsquigarrow F \otimes G$ is a binary operation on the set Π. We will denote it by *T-convolution* or, whenever there is no danger of confusion, simply *convolution*. It may easily be seen that it is an operation which is commutative and distributive with respect to addition; moreover, for $F, G \in \Pi$ and $\alpha \in \mathbb{C}$

$$\alpha(F \otimes G) \ = \ (\alpha F) \otimes G \ = \ F \otimes (\alpha G).$$

To verify that Π is in fact an algebra it will therefore be enough to prove the associativity of the convolution.

Let $F, G, H \in \Pi$ and denote by A the square subset of $\mathbb{R}_u \times \mathbb{R}_v$ defined by the relations:

$$0 \leq u \leq T, \qquad 0 \leq v \leq T.$$

For any $x \in \mathbb{R}$, we will have:

$$(F \otimes G) \otimes H(x) \ = \ \int_0^T \left(\int_0^T F(v - u)G(u) \, du \right) H(x - v) \, dv$$

$$= \ \int\!\!\int_A F(v - u)G(u)H(x - v) \, du \, dv.$$

This integral is transformed by the change of variables

$$\begin{cases} u = x - \xi - \eta \\ v = x - \eta \end{cases}$$

into

$$\iint_{D^*} F(\xi)G(x - \xi - \eta)H(\eta)\, d\xi d\eta, \tag{3.22}$$

where D^* is the parallelogram in $\mathbb{R}_\xi \times \mathbb{R}_\eta$ defined by

$$x - T \leq \eta \leq x, \qquad x - T \leq \xi + \eta \leq x.$$

Denote by B^* and C^*, respectively, the intersections of D^* with the half-planes $\xi > 0$ and $\xi < 0$ and by C_1^* the triangle that results from C^* by a translation of amplitude T in the direction and sense of the ξ-axis. Then taking into account the fact that for each fixed η the integrand is periodic in ξ with period T, it is easy to see that the integral (3.22) is equal to

$$\iint_{B^* \cup C_1^*} F(\xi)G(x - \xi - \eta)H(\eta)\, d\xi d\eta.$$

On the other hand, since the integrand is periodic in η for each ξ, this last integral has the same value as

$$\iint_{A^*} F(\xi)G(x - \xi - \eta)H(\eta)\, d\xi d\eta,$$

where A^* is the square $[0, T] \times [0, T]$ in the (ξ, η)-plane. It follows then that

$$\begin{aligned} (F \otimes G) \otimes H(x) &= \iint_{A^*} F(\xi)G(x - \xi - \eta)H(\eta)\, d\xi d\eta \\ &= \int_0^T F(\xi) \left(\int_0^T G(x - \xi - \eta)H(\eta)\, d\eta \right) d\xi \\ &= F \otimes (G \otimes H)(x). \end{aligned}$$

Here are some simple examples of convolutions in the space Π^T:

1. If $F \in \Pi^T$ then

$$\mathbf{1} \otimes F(x) = \int_0^T F(t)\, dt,$$

2. For $\omega = \frac{2\pi}{T}$ we get

$$\begin{aligned} e^{im\omega x} \otimes e^{in\omega x} &= e^{im\omega x} \int_0^T e^{i(n-m)\omega t}\, dt \\ &= \begin{cases} 0 & \text{if } m \neq n \\ T e^{im\omega x} & \text{if } m = n. \end{cases} \end{aligned}$$

Another result which will be of interest in the sequel is the following

Proposition 3.21 *Let $(F_n)_{n\in\mathbb{N}}$ be a sequence of continuous periodic functions uniformly convergent (in $\mathbb{R}$) to a (continuous periodic) function F. Then, if G is another continuous periodic function, we have*

$$F \otimes G(x) \;=\; \lim_{n\to\infty} \left[F_n \otimes G(x) \right],$$

where the convergence is uniform in $\mathbb{R}$.

Proof: Given $\varepsilon > 0$, set $N = \max_{t\in\mathbb{R}} |G(t)|$. We may assume that $N \neq 0$ since otherwise the result is trivial. Hence we may choose p such that for $n > p$ and $x, t \in \mathbb{R}$ we have

$$|F_n(x - t) - F(x - t)| \;<\; \frac{\varepsilon}{NT}.$$

Thus, for any $x \in \mathbb{R}$ we will also have for any $n > p$:

$$\begin{aligned} |F_n \otimes G(x) - F \otimes G(x)| \;&\leq\; \int_0^T |F_n(x - t) - F(x - t)|\,|G(t)|\,dt \\ &\leq\; T\,\frac{\varepsilon}{NT}\,N \;=\; \varepsilon \end{aligned}$$

from which the proposition follows. $\qquad\square$

In order to extend the convolution to periodic distributions, it is convenient to introduce now the following proposition:

Proposition 3.22 *Let $F, G \in \Pi$ and suppose that $F \in \mathcal{C}^1(\mathbb{R})$. Then $F \otimes G \in \mathcal{C}^1(\mathbb{R})$ and, moreover $(F \otimes G)' = F' \otimes G$.*

Proof: The proof is similar to the proof of proposition 2.31, but simpler. Given $\varepsilon > 0$, choose $\gamma > 0$ such that, for $|h| < \gamma$, we have, for any $x \in \mathbb{R}$:

$$|F'(x + h) - F'(x)| \;<\; \frac{\varepsilon}{NT}$$

where $N = \max_{x\in\mathbb{R}} |G(x)|$ (which, as usual, and without any loss of generality, we may suppose different from zero). Then, taking into account that, under the conditions stated in the hypothesis

$$F(x - t + h) - F(x - t) \;=\; hF'(x - t + \theta h),$$

with $\theta \in\,]0, 1[$, we will have for $x \in \mathbb{R}$ and $|h| < \gamma$:

$$\left| \frac{1}{h}[F \otimes G(x + h) - F \otimes G(x)] - F' \otimes G(x) \right|$$

$$= \left| \int_0^T [F'(x - t + \theta h) - F'(x - t)]G(t)\,dt \right| \;\leq\; \frac{\varepsilon}{NT}\,NT \;=\; \varepsilon,$$

from which the proposition follows. $\qquad\square$

The following corollaries of this proposition are then easy to obtain

Corollary 3.23 *If $F, G \in \Pi$, $F \in \mathcal{C}^p(\mathbb{R})$ and $G \in \mathcal{C}^q(\mathbb{R})$, then $F \otimes G$ belongs to $\mathcal{C}^{p+q}(\mathbb{R})$, and moreover $(F \otimes G)^{(p+q)} = F^{(p)} \otimes G^{(q)}$.*

Corollary 3.24 *If $F, G \in \Pi$, $F \in \mathcal{C}^\infty(\mathbb{R})$ then $F \otimes G \in \mathcal{C}^\infty(\mathbb{R})$ and, for each $p \in \mathbb{N}$, $(F \otimes G)^{(p)} = F^{(p)} \otimes G$.*

Denote by Π_∞^T (or, simply, by Π_∞) the subspace of $\mathcal{C}_\infty(\mathbb{R})$ comprising all distributions that admit the period T, that is, that satisfy the condition

$$\tau_T f(\hat{x}) \; = \; f(\hat{x} - T) \; = \; f(\hat{x}).$$

Whenever there is no danger of confusion we shall shorten the expression *"distribution which admits the period T"* to *"periodic distribution"*.

Let f be a periodic distribution and g an arbitrary primitive of f; we will then have:

$$\mathbf{D}[g(\hat{x}) - g(\hat{x} - T)] \; = \; f(\hat{x}) - f(\hat{x} - T) \; = \; 0,$$

which shows that the difference between the distributions $g(\hat{x})$ and $g(\hat{x} - T)$ is a constant (which depends upon f, but is independent of the primitive g considered). We will call this constant *the integral of f over the period interval* and will denote it by the symbol

$$\int_T f(x) \, dx$$

eventually abbreviated to $\int_T f$; by definition we will therefore have

$$\int_T f(x) \, dx \; = \; g(\hat{x}) - g(\hat{x} - T) \; = \; g - \tau_T g.$$

If f_1 and f_2 are two periodic distributions and $\alpha_1, \alpha_2 \in \mathbf{C}$ then clearly we have

$$\int_T (\alpha_1 f_1 + \alpha_2 f_2) \; = \; \alpha_1 \int_T f_1 + \alpha_2 \int_T f_2.$$

In the particular case when f is a continuous periodic function, it is obvious that

$$\int_T f(x) \, dx \; = \; \int_0^T f(x) \, dx.$$

If f is a periodic distribution which is precontinuous at a point a (and, therefore, also in $a + T$) we also have

$$\int_T f(x) \, dx \; = \; \int_a^{a+T} f(x) \, dx,$$

where the integral of the right-hand side is to be interpreted in the sense which has been defined in section 2.5.

A very simple property of the integral is that one which is expressed by the following *"formula of integration by parts"*:

Proposition 3.25 *Let $\varphi \in \mathcal{C}^\infty(\mathbb{R}) \cap \Pi^T$ and $f \in \Pi_\infty^T$; then we have*

$$\int_T \varphi \mathbf{D} f \; = \; - \int_T \varphi' f.$$

More generally, for any $p \in \mathbb{N}$, we have

$$\int_T \varphi \mathbf{D}^p f \;=\; (-1)^p \int_T \varphi^{(p)} f.$$

Proof: The first formula is immediately obtained by integrating both sides of the equality:

$$\mathbf{D}(\varphi f) \;=\; \varphi \mathbf{D} f + \varphi' f$$

and taking into account the periodicity of φf; the second formula is an immediate consequence of the first one. $\qquad\square$

If f is a periodic distribution, we call the number defined by

$$\mu(f) \;=\; \frac{1}{T} \int_T f(x)\,dx$$

the mean value of the distribution f.[11] It is easy to see that the map $\mu : \Pi_\infty \to \mathbb{C}$ defined in this way is linear and also that, for a constant distribution, $f(x) = c$, we have $\mu(f) = c$.

As examples of periodic distributions with period T we will mention the derivatives (of order ≥ 1) of the function $C(\frac{x}{T})$ (where, as usual, $C(\frac{x}{T})$ denotes the largest integer less than or equal to $\frac{x}{T}$); for each positive integer p, the restriction of $\mathbf{D}^p C(\frac{x}{T})$ to the interval $]-T, +T[$ coincides with the restriction to the same interval of the distribution $\delta^{(p-1)}$.

The distribution $\mathbf{D}C(\frac{x}{T})$ is usually called the *"periodic δ"* (with period T); to denote such a distribution we will use the symbol $\tilde{\delta}_T$; the derivatives of this distribution will be denoted by $\tilde{\delta}'_T$, $\tilde{\delta}''_T$, etc.

From the relation

$$C\left(\frac{x}{T}\right) - C\left(\frac{x-T}{T}\right) \;=\; 1$$

it follows that the integral of $\tilde{\delta}_T$ over the period interval is equal to 1; we therefore have that $\mu\left(\tilde{\delta}_T\right) = \frac{1}{T}$. For the derivatives $\tilde{\delta}'_T$, $\tilde{\delta}''_T$, etc., as for any other distribution which is the derivative of a periodic distribution, the mean value is zero.

We leave the proof of the following proposition to the reader:

Proposition 3.26 *For a periodic distribution f to have a (first) primitive which is periodic (and therefore for all first primitives to be periodic) it is necessary and sufficient that its mean value be zero. If this condition is satisfied f will have just one primitive with zero mean value.*

---

[11]It is not difficult to see that if T^* is any other period of the distribution f then it will necessarily be true that

$$\frac{1}{T^*} \int_{T^*} f(x)\,dx \;=\; \frac{1}{T} \int_T f(x)\,dx,$$

and therefore that the mean value of the periodic distribution f is independent of the period considered.

From this property it follows immediately that the operator $\mathbf{D}$, restricted to the subspace of Π_∞ comprising all distributions which have zero mean value (which we will denote by Π^0_∞), is invertible; hence, $\mathbf{D}$ may be viewed as an automorphism on Π^0_∞.

Clearly, the space Π^0 of continuous periodic functions with zero mean value (in the usual sense) is a linear subspace of Π^0_∞; and it is easy to see that:

> **Proposition 3.27** *If f is a periodic distribution with zero mean value, there exist $p \in \mathbb{N}$ and $F \in \Pi^0$, a periodic continuous function with zero mean value, such that $f = \mathbf{D}^p F$.*

Before continuing we will prove the following result, which will be needed in the sequel:

> **Proposition 3.28** *Let f and g be periodic distributions (with period T); if there exists an interval $\mathbf{I}$, with length greater than T, such that $f = g$ on $\mathbf{I}$, then $f = g$.*

Proof: It is obviously enough to prove that, if the periodic distribution f is null on an interval $\mathbf{I}$, with length greater than T, then $f = \mathbf{0}$.

We begin by showing that, under the given hypothesis, we have $\mu(f) = 0$. To this end set

$$\mathbf{J} = \{x : x \in \mathbf{I} \text{ and } x - T \in \mathbf{I}\}$$

and denote by g a (first) primitive of f.

Since $g(x)$ is constant on $\mathbf{I}$, the distribution $g(x) - g(x - T)$ will be null on $\mathbf{J}$; hence, restricting to this interval both members of the equality

$$T\mu(f) = g(\hat{x}) - g(\hat{x} - T),$$

one sees immediately that $\mu(f) = 0$.

Therefore we have that $f \in \Pi^0_\infty$ and from the previous proposition it follows that there exist a natural number p and a continuous periodic function F such that $f = \mathbf{D}^p F$ on $\mathbb{R}$. Taking into account again the fact that $f = \mathbf{0}$ on $\mathbf{I}$, it may be deduced that there exists a polynomial $P(x)$, of degree $< p$, such that $F(x) = P(x)$ on $\mathbf{I}$.

Since $F(x) - F(x - T) = 0$ for every $x \in \mathbb{R}$, it must also be true that $P(x) - P(x - T) = 0$ for each $x \in \mathbf{J}$, from which it follows that the polynomial $P(x) - P(x - T)$ with an infinity of real roots must be identically zero. Thus since both F and P are periodic functions, the equality $F(x) = P(x)$ must be satisfied at every point $x \in \mathbb{R}$, which allows us to conclude that $f = \mathbf{D}^p F = \mathbf{0}$. $\square$

The proposition 3.29 may now be stated, the proof being similar to that of proposition 2.34:

> **Proposition 3.29** *There exists one and only one mapping $(f, g) \rightsquigarrow f \otimes g$, from $\Pi^0_\infty \times \Pi^0_\infty$ into Π^0_∞ satisfying the following conditions:*
>
> i) *if f and g are continuous functions which are periodic with zero mean value, then $f \otimes g$ is the T-convolution of f and g defined above;*

ii) *for any $f, g \in \Pi_\infty^0$,*

$$\mathbf{D}(f \otimes g) \;=\; (\mathbf{D}f) \otimes g \;=\; f \otimes (\mathbf{D}g).$$

The distribution $f \otimes g$ referred to in the above proposition may clearly be defined by the formula:

$$f \otimes g \;=\; \mathbf{D}^{p+q}(F \otimes G),$$

where $F, G \in \Pi^0$ and $f = \mathbf{D}^p F$, $g = \mathbf{D}^q G$. It should be noted that the convolution of two periodic continuous functions has zero mean value whenever the same holds for either of the convolved distributions.

Example 3.30 Let F be a function in Π^0 and G the periodic extension of the function G^* defined on $[0, T]$ by the formula:

$$G^*(x) \;=\; -\frac{1}{2T}\, x^2 + \frac{1}{2}\, x - \frac{T}{12}.$$

It is easy to see that we have $G \in \Pi^0$. Denoting by F_1 and F_2 the elements in Π^0 such that

$$\mathbf{D}^2 F_2 \;=\; \mathbf{D}F_1 \;=\; F,$$

we will have, for any $x \in \mathbb{R}$:

$$
\begin{aligned}
F \otimes G(x) \;&=\; \int_0^T G(t)F(x - t)\, dt \\
&=\; -\frac{T}{12}\int_0^T F(x - t)\, dt + \frac{1}{2}\int_0^T tF(x - t)\, dt - \frac{1}{2T}\int_0^T t^2 F(x - t)\, dt.
\end{aligned}
$$

The first integral on the right-hand side is zero, since $F \in \Pi^0$; by integrating by parts the two following integrals, we obtain

$$F \otimes G(x) \;=\; \frac{1}{2}\left\{ [-tF_1(x - t)]_0^T + \int_0^T F_1(x - t)\, dt \right\}$$

$$-\frac{1}{2T}\left\{ [-t^2 F_1(x - t)]_0^T - [2t F_2(x - t)]_0^T + 2\int_0^T F_2(x - t)\, dt \right\}.$$

Since both integrals in this formula vanish, we may conclude that

$$F \otimes G(x) \;=\; -\frac{T}{2} F_1(x - T) - \frac{1}{2T}\left[-T^2 F_1(x - T) - 2T F_2(x - T) \right] \;=\; F_2(x).$$

Now let f be a periodic distribution with zero mean value but otherwise arbitrary; representing f in the form $f = \mathbf{D}^p F$, with $F \in \Pi^0$, and using the fact that, as easily seen, we have

$$\mathbf{D}^2 G \;=\; \tilde{\delta}_T - \frac{1}{T}$$

it follows that

$$f \otimes \left(\tilde{\delta}_T - \frac{1}{T} \right) \;=\; \mathbf{D}^{p+2}(F \otimes G) \;=\; \mathbf{D}^{p+2} F_2 \;=\; \mathbf{D}^p F \;=\; f.$$

We can now define the convolution of two arbitrary periodic distributions. If f, g are in Π_∞, set $\alpha = \mu(f)$, $\beta = \mu(g)$, $f_0 = f - \alpha$ and $g_0 = g - \beta$. The convolution in the space Π_∞ can then be defined by the formula

$$f \otimes g \;=\; T\alpha\beta + f_0 \otimes g_0,$$

where $f_0 \otimes g_0$ is the convolution of f_0 and g_0 in the space Π_∞^0 according to proposition 3.29. (We remark that in the particular case when f and g have zero mean value, the convolution of f and g in Π_∞ coincides with the convolution of the same distributions in the space Π_∞^0).

We have now

Proposition 3.31 *With the convolution product defined as above, the space Π_∞ of periodic distributions with its usual linear space structure, is a complex commutative algebra where $\tilde{\delta}_T$ is the unit element.*

Proof: We leave to the reader the proof that the convolution is associative, commutative and distributive with respect to addition and, moreover, that for $f, g \in \Pi_\infty$ and $\alpha \in \mathbf{C}$, we have

$$\alpha(f \otimes g) \;=\; (\alpha f) \otimes g \;=\; f \otimes (\alpha g).$$

To see that $\tilde{\delta}_T$ is the unit element of the convolution algebra Π_∞ it is enough to note that if $f \in \Pi_\infty$, $\mu(f) = \alpha$ and $f_0 = f - \alpha$ we will have (appealing to the preceding examples)

$$\begin{aligned}
\tilde{\delta}_T \otimes f \;&=\; \left[\frac{1}{T} + \left(\tilde{\delta}_T - \frac{1}{T} \right) \right] \otimes (\alpha + f_0) \\
&=\; T\frac{1}{T}\alpha + \left(\tilde{\delta}_T - \frac{1}{T} \right) \otimes f_0 \;=\; \alpha + f_0 \;=\; f,
\end{aligned}$$

and the proof is complete. $\square$

Proposition 3.32 *For $f, g \in \Pi_\infty$ and $h \in \mathbb{R}$ we have*

i) $\mathbf{D}(f \otimes g) \;=\; (\mathbf{D}f) \otimes g \;=\; f \otimes (\mathbf{D}g)$;

ii) $\tau_h(f \otimes g) \;=\; (\tau_h f) \otimes g \;=\; f \otimes (\tau_h g)$.

Proof: Taking into account the commutativity of the convolution it will be enough, in each case, to prove only the first equality. Accordingly, let $f = \alpha + f_0$ and $g = \beta + g_0$, with $f_0, g_0 \in \Pi_\infty^0$ such that $f_0 = \mathbf{D}^p F_0$, $g_0 = \mathbf{D}^q G_0$, where $F_0, G_0 \in \Pi^0$. Then

$$\begin{aligned}
\mathbf{D}(f \otimes g) \;&=\; \mathbf{D}\left(T\alpha\beta + f_0 \otimes g_0 \right) \;=\; \mathbf{D}^{p+q+1}\left(F_0 \otimes G_0 \right) \\
&=\; \mathbf{D}f_0 \otimes g_0 \;=\; \mathbf{D}f_0 \otimes (\beta + g_0) \;=\; \mathbf{D}f \otimes g.
\end{aligned}$$

Then, taking into account in particular the relations

$$\tau_h\left(F_0 \otimes G_0 \right) \;=\; (\tau_h F_0) \otimes G_0$$

and

$$\mu\left(\tau_h f \right) \;=\; \mu(f)$$

which are easily verified, one obtains immediately:

$$
\begin{aligned}
\tau_h(f \otimes g) &= \tau_h\left(T\alpha\beta + f_0 \otimes g_0\right) \\
&= T\alpha\beta + \tau_h\left[\mathbf{D}^{p+q}\left(F_0 \otimes G_0\right)\right] = T\alpha\beta + \mathbf{D}^{p+q}\tau_h\left(F_0 \otimes G_0\right) \\
&= T\alpha\beta + \mathbf{D}^{p+q}\left[(\tau_h F_0) \otimes G_0\right] = T\alpha\beta + (\tau_h f_0) \otimes g_0 = (\tau_h f) \otimes g,
\end{aligned}
$$

as was to be proved. $\qquad\qquad\square$

From these properties it follows that, for any periodic distribution f and any $p \in \mathbb{N}$ and $h \in \mathbb{R}$, we have

$$
\tilde{\delta}_T^{(p)} \otimes f \;=\; \left(\mathbf{D}^p \tilde{\delta}_T\right) \otimes f \;=\; \mathbf{D}^p\left(\tilde{\delta}_T \otimes f\right) \;=\; \mathbf{D}^p f
$$

and

$$
\tilde{\delta}_T(\hat{x} - h) \otimes f \;=\; \left(\tau_h \tilde{\delta}_T\right) \otimes f \;=\; \tau_h\left(\tilde{\delta}_T \otimes f\right) \;=\; \tau_h f.
$$

In the same way we also have

$$
\tilde{\delta}_T^{(p)}(\hat{x} - h) \otimes f \;=\; \left(\mathbf{D}^p \tau_h \tilde{\delta}_T\right) \otimes f \;=\; \mathbf{D}^p f(\hat{x} - h).
$$

Another property of the convolution in the algebra Π_∞ which is important in the sequel is the following:

Proposition 3.33 *If φ is a C^∞-periodic function and f is a periodic distribution, $\varphi \otimes f$ is a (periodic) function of class C^∞ and at each point $x \in \mathbb{R}$,*

$$
\varphi \otimes f(x) \;=\; \int_T \varphi(x - t) f(t)\, dt.
$$

Proof: Let $f = \alpha + f_0$, with $f_0 = \mathbf{D}^p F_0$ and $F_0 \in \Pi^0$, and let $\varphi = \beta + \varphi_0$, with $\beta = \mu(\varphi)$. Then

$$
\varphi \otimes f \;=\; T\alpha\beta + \mathbf{D}^p(\varphi_0 \otimes F_0)
$$

and, since $\varphi_0 \otimes F_0$ is a function of class C^∞ (by the second corollary of the proposition 3.22), the same will hold for $\varphi \otimes f$.

On the other hand

$$
\varphi \otimes f(x) \;=\; T\alpha\beta + \int_T \varphi_0^{(p)}(x - t) F_0(t)\, dt,
$$

from which, by integrating by parts, and using the fact that

$$
\int_T \beta f_0(t)\, dt \;=\; 0,
$$

it follows that

$$
\varphi \otimes f(x) \;=\; T\alpha\beta + \int_T \varphi_0(x - t) f_0(t)\, dt \;=\; T\alpha\beta + \int_T \varphi(x - t) f_0(t)\, dt,
$$

and, finally, that

$$
\varphi \otimes f(x) \;=\; \int_T \alpha\varphi(x - t)\, dt + \int_T \varphi(x - t) f_0(t)\, dt \;=\; \int_T \varphi(x - t) f(t)\, dt
$$

as was to be shown.   □

We have already had occasion to remark that the operator $\mathbf{D}$, restricted to the space Π^0_∞, is invertible; denoting its inverse (defined on Π^0_∞) by $\mathbf{D}^{-1}$, we may state the proposition:

Proposition 3.34 *Let $(f_n)_{n\in\mathbf{N}}$ be a sequence converging to f in $\mathcal{C}_\infty(\mathbb{R})$. Then*

i) *if the distributions f_n belong to Π_∞, the distribution f is in Π_∞, and*

$$\lim_{n\to\infty} \mu(f_n) \;=\; \mu(f);$$

ii) *if the distributions f_n belong to Π^0_∞, the distribution f is in Π^0_∞ and the sequence $(\mathbf{D}^{-1}f_n)_{n\in\mathbf{N}}$ converges in $\mathcal{C}_\infty(\mathbb{R})$ to the distribution $\mathbf{D}^{-1}f$.*

Proof: i) Since the limit operation commutes with translations (proposition 3.5), from $f_n(\hat{x} - T) = f_n(\hat{x})$ it follows at once that $f(\hat{x} - T) = f(\hat{x})$; that is, $f \in \Pi_\infty$.

On the other hand, appealing to proposition 3.13, there exists a convergent sequence $(g_n)_{n\in\mathbf{N}}$ such that $\mathbf{D}g_n = f_n$ for each n. Setting $g = \lim_{n\to\infty} g_n$ we will have that $\mathbf{D}g = \lim_{n\to\infty} \mathbf{D}g_n = f$, and

$$\lim_{n\to\infty} \mu(f_n) \;=\; \frac{1}{T}\lim_{n\to\infty}[g_n(\hat{x}) - g_n(\hat{x} - T)] \;=\; \frac{1}{T}[g(\hat{x}) - g(\hat{x} - T)] \;=\; \mu(f).$$

ii) From what has just been proved, if for each $n \in \mathbf{N}$ we have $f_n \in \Pi^0_\infty$ (that is, $\mu(f_n) = 0$) then we also have $\mu(f) = 0$ and, therefore, $f \in \Pi^0_\infty$. On the other hand, if we choose a convergent sequence $(g_n)_{n\in\mathbf{N}}$ such that $\mathbf{D}g_n = f_n$ (cf. proposition 3.13) and if we set $g = \lim_{n\to\infty} g_n$, then g and g_n will be periodic distributions (in view of the fact that they are primitives of periodic distributions with zero mean value) and we will have

$$\lim_{n\to\infty}[g_n - \mu(g_n)] \;=\; g - \lim_{n\to\infty}\mu(g_n) \;=\; g - \mu(g).$$

Since obviously $g_n - \mu(g_n) = \mathbf{D}^{-1}f_n$ and $g - \mu(g) = \mathbf{D}^{-1}f$, we may consider the proof complete.   □

Proposition 3.35 *Let $(f_n)_{n\in\mathbf{N}}$ be a sequence of distributions in Π^0_∞ which converges in $\mathcal{C}_\infty(\mathbb{R})$ to the distribution f. Then there exists $p \in \mathbf{N}$ and $F, F_n \in \Pi^0$ such that $f = \mathbf{D}^p F$ and $f_n = \mathbf{D}^p F_n$ where $(F_n)_{n\in\mathbf{N}}$ converges to F uniformly on $\mathbb{R}$.*

Proof: Given a compact interval K, with length greater than T, there exist $p \in \mathbf{N}$ and $G, G_n \in \mathcal{C}(K)$ such that $(G_n)_{n\in\mathbf{N}}$ converges to G uniformly on K and $f = \mathbf{D}^p G$, $f_n = \mathbf{D}^p G_n$. Then set (in $\mathbb{R}$)

$$F_n \;=\; \mathbf{D}^{-p}f_n, \quad F \;=\; \mathbf{D}^{-p}f.$$

By the previous proposition, the sequence $(F_n)_{n\in\mathbf{N}}$ will converge to F in $\mathcal{C}_\infty(\mathbb{R})$. On the other hand, since $\mathbf{D}^p G_n = \mathbf{D}^p F_n$ on K there will exist polynomials P_n of degree $< p$ such that

$$F_n(x) - G_n(x) \;=\; P_n(x),$$

for each $x \in K$. Since $(F_n)_{n \in \mathbb{N}}$ and $(G_n)_{n \in \mathbb{N}}$ both converge in $\mathcal{C}_\infty(K)$ the same will be true with respect to the sequence $(P_n)_{n \in \mathbb{N}}$ which converges to a limit P. P will be a polynomial of degree $< p$ and the convergence of P_n to P will be uniform on K (cf. the proof of proposition 3.11).

Therefore, using the fact that the convergence of $(G_n)_{n \in \mathbb{N}}$ to G is also uniform on K, it follows that the same will be true with respect to the convergence of $(F_n)_{n \in \mathbb{N}}$ to F; and since F_n and F are periodic functions and the interval K has length greater than the period T it may be concluded that $(F_n)_{n \in \mathbb{N}}$ converges uniformly in $\mathbb{R}$. $\qquad\square$

We may now state the theorem:

Theorem 3.36 *Let $(f_n)_{n \in \mathbb{N}}$ be a convergent sequence of periodic distributions and $f = \lim_{n \to \infty} f_n$; also let g be a periodic distribution. Under these conditions we have $\lim_{n \to \infty}(f_n \otimes g) = f \otimes g$.*

Proof: Consider first the case when all the distributions g and f_n (and therefore also f) have zero mean value. According to propositions 3.35 and 3.27 it follows that there exist integers p, q and functions $G, F, F_n \in \Pi^0$ such that $f = \mathbf{D}^p F$, $f_n = \mathbf{D}^p F_n$ and $g = \mathbf{D}^q G$, where $(F_n)_{n \in \mathbb{N}}$ converges to F uniformly on $\mathbb{R}$. The proposition 3.21 therefore shows that the sequence $(F_n \otimes G)_{n \in \mathbb{N}}$ converges to $F \otimes G$ uniformly on $\mathbb{R}$ and thus we have

$$\lim_{n \to \infty}(f_n \otimes g) = \lim_{n \to \infty} \mathbf{D}^{p+q}(F_n \otimes G)$$
$$= \mathbf{D}^{p+q} \lim_{n \to \infty}(F_n \otimes G) = \mathbf{D}^{p+q}(F \otimes G) = f \otimes g.$$

In the general case, with $f, g \in \Pi_\infty$ and $f = \lim_{n \to \infty} f_n$, set $f_n = \alpha_n + f_{0,n}$, $f = \alpha + f_0$, $g = \beta + g_0$, with $f_{0,n}, f_0, g_0 \in \Pi^0_\infty$. We will then have

$$\alpha = \mu(f) = \lim_{n \to \infty} \mu(\alpha_n) = \lim_{n \to \infty} \alpha_n,$$

and

$$f_0 = f - \alpha = \lim_{n \to \infty} f_n - \lim_{n \to \infty} \alpha_n = \lim_{n \to \infty} f_{0,n},$$

and, therefore, taking into account what we have just proved,

$$\lim_{n \to \infty}(f_n \otimes g) = \lim_{n \to \infty}(T\alpha_n\beta + f_{0,n} \otimes g_0)$$
$$= T\alpha\beta + \lim_{n \to \infty}(f_{0,n} \otimes g_0) = T\alpha\beta + f_0 \otimes g_0 = f \otimes g.$$

as asserted. $\qquad\square$

The theory of Fourier series, which we will start to approach now, is particularly simple in the setting of distribution theory. We have already proved, in section 3.2, that a sufficient condition for the convergence of the series

$$\sum_{n=-\infty}^{+\infty} c_n e^{in\omega x} \qquad (\omega > 0) \tag{3.23}$$

in $\mathcal{C}_\infty(\mathbb{R})$ is that the coefficients c_n constitute a tempered sequence, that is, that there exist constants M and p such that, for every $n \neq 0$, we have

$$|c_n| \leq M |n|^p.$$

It is clear that whenever the series converges, its sum is a periodic distribution, with period $T = 2\pi/\omega$ (it is enough to note that each one of the terms of the series admits that period and that the sum and limit operations commute with translations).

Further, if the sum of the series (3.23) is known in advance the coefficients c_n may be easily computed. In fact, if the equality

$$f(x) = \sum_{n=-\infty}^{+\infty} c_n e^{in\omega x}$$

holds, we have for any $k \in \mathbb{Z}$ (using the interchangeability of multiplication with the limit operation):

$$e^{-ik\omega x} f(x) = \sum_{n=-\infty}^{+\infty} c_n e^{i(n-k)\omega x}$$

from which, by the linearity and continuity of the mean value we get

$$\mu\left(e^{-ik\omega x} f(x)\right) = c_k,$$

or

$$c_k = \frac{1}{T} \int_T e^{-ik\omega x} f(x) \, dx. \tag{3.24}$$

Since the product of the periodic function $e^{-ik\omega x}$ by a periodic distribution is still a periodic distribution, the coefficients c_k defined by the formula (3.24) may be computed for any periodic distribution f, without needing the prior assumption that f is represented by a series of the form (3.23). These are the so-called *Fourier coefficients* of the given periodic distribution f.

For the distribution $\tilde{\delta}_T$, for example, we have for any $n \in \mathbb{Z}$,

$$e^{-in\omega x} \tilde{\delta}_T(x) = \tilde{\delta}_T(x)$$

which follows easily from the fact that the left-hand side is a periodic distribution whose restriction to the interval $]-T, +T[$ is identical with the distribution $e^{-in\omega x}\delta(x) = \delta(x)$. Hence the Fourier coefficients of $\tilde{\delta}_T$,

$$c_n = \frac{1}{T} \int_T e^{-in\omega x} \tilde{\delta}_T(x) \, dx = \frac{1}{T} \int_T \tilde{\delta}_T(x) \, dx = \frac{1}{T}$$

are all equal to $1/T$.

We will now show that the Fourier coefficients of any periodic distribution

$$c_n = \frac{1}{T} \int_T e^{-in\omega x} f(x) \, dx$$

constitute a tempered sequence, and therefore that the series

$$\sum_{n=-\infty}^{+\infty} c_n e^{in\omega x}$$

called the *Fourier series* of f, is always convergent. In particular it then follows that a series of the form (3.23) can only converge if its coefficients c_n constitute a tempered sequence, that is, that the condition of convergence which we have proved to be sufficient is also necessary. Finally we will also prove that the sum of the Fourier series of an arbitrary periodic distribution is that distribution itself.

Thus, let f be a periodic distribution, say $f = \alpha + \mathbf{D}^p F_0$, with $F_0 \in \Pi^0$, and let

$$c_n = \frac{1}{T} \int_T e^{-in\omega x} f(x)\, dx$$

be its n-th Fourier coefficient. Integrating by parts, we obtain for any $n \neq 0$

$$\begin{aligned}
c_n &= \frac{1}{T} \int_T \alpha e^{-in\omega x} dx + \frac{1}{T} \int_T e^{-in\omega x} \mathbf{D}^p F_0(x)\, dx \\
&= \frac{(-1)^p}{T} (-in\omega)^p \int_T e^{-in\omega x} F_0(x)\, dx
\end{aligned}$$

and so

$$|c_n| \leq \frac{\omega^p n^p}{T} \int_0^T |F_0(x)|\, dx,$$

which proves the following result

Proposition 3.37 *The Fourier coefficients of any periodic distribution f constitute a tempered sequence.*

Thus we have that for any periodic distribution f the convergence of its corresponding Fourier series is assured. To prove that the sum of this series is precisely the distribution f, it is convenient to start with the particular case of $f = \tilde{\delta}_T$:

Proposition 3.38 *The Fourier series for $\tilde{\delta}_T$,*

$$\sum_{n=-\infty}^{+\infty} \frac{1}{T} e^{in\omega x}$$

has $\tilde{\delta}_T$ as its sum.

Proof: Since the sequence of the Fourier coefficients, $c_n = 1/T$, is tempered then the series is convergent. Denote by γ its sum, which is certainly a periodic distribution.

Denoting by $\mathbf{I}$ the interval $]-T, +T[$, we begin by proving that there exists $c \in \mathbf{C}$ such that $\gamma = c\delta$ on $\mathbf{I}$. For this note first that we have

$$e^{i\omega x} \gamma(x) = \sum_{n=-\infty}^{+\infty} \frac{1}{T} e^{i(n+1)\omega x} = \sum_{n=-\infty}^{+\infty} \frac{1}{T} e^{in\omega x} = \gamma(x),$$

that is,

$$\left(e^{i\omega x} - 1 \right) \gamma(x) = 0. \tag{3.25}$$

Since $e^{i\omega x} - 1$ is different from zero in each one of the intervals $]-T, 0[$ and $]0, +T[$, it follows that γ will be null on both of these intervals and therefore that there must exist complex numbers α_j, $j = 0, 1, \ldots, m$ such that we have

$$\gamma(x) \;=\; \sum_{j=0}^{m} \alpha_j \delta^{(j)}(x)$$

on **I**. On the other hand we have

$$e^{i\omega x} \;=\; 1 + x\theta(x),$$

where θ is a function which is different from zero on the interval **I**; hence, in this interval the equality (3.25) is equivalent to $x\gamma(x) = 0$ or

$$x \sum_{j=0}^{m} \alpha_j \delta^{(j)}(x) \;=\; 0$$

or, again,

$$\sum_{j=1}^{m} (-j\alpha_j) \delta^{(j-1)}(x) \;=\; 0.$$

Thus $\alpha_j = 0$, $(j = 1, 2, \ldots, m)$, that is, there exists $c \in \mathbf{C}$ such that $\gamma = c\delta$ on **I**. From proposition 3.28 it then follows that we must have

$$\gamma \;=\; c\tilde{\delta}_T$$

(in $\mathbb{R}$). To see that $c = 1$ it is enough to note that, on the one hand

$$\mu(\gamma) \;=\; \mu(c\,\tilde{\delta}_T) \;=\; c\,\mu(\tilde{\delta}_T) \;=\; \frac{c}{T}$$

and on the other,

$$\mu(\gamma) \;=\; \mu\left(\sum_{n=-\infty}^{+\infty} \frac{1}{T}\, e^{in\omega x} \right) \;=\; \sum_{n=-\infty}^{+\infty} \frac{1}{T}\, \mu(e^{in\omega x}) \;=\; \frac{1}{T}$$

Then

$$\tilde{\delta}_T(x) \;=\; \sum_{n=-\infty}^{+\infty} \frac{1}{T}\, e^{in\omega x}$$

as was to be proved. $\square$

Proposition 3.39 *The Fourier series of a periodic distribution f has the distribution f as its sum.*

Proof: If f is a periodic distribution, the proposition 3.32 allows us to see that we must have

$$e^{in\omega x} \otimes f(x) \;=\; \int_T e^{in\omega(x-t)} f(t)\, dt$$

$$\;=\; e^{in\omega x} \int_T e^{-in\omega t} f(t)\, dt \;=\; T c_n e^{in\omega x}$$

where c_n denotes the Fourier coefficient of order n of the distribution f. Hence, appealing to the propositions 3.29, 3.35 and 3.37 it follows immediately that

$$
\begin{aligned}
f(x) &= f(x) \otimes \tilde{\delta}_T(x) = f(x) \otimes \sum_{n=-\infty}^{+\infty} \frac{1}{T} e^{in\omega x} \\
&= \sum_{n=-\infty}^{+\infty} \frac{1}{T} e^{in\omega x} \otimes f(x) = \sum_{n=-\infty}^{+\infty} \frac{1}{T} T c_n e^{in\omega x} = \sum_{n=-\infty}^{+\infty} c_n \, e^{in\omega x}
\end{aligned}
$$

and the result follows. $\qquad\square$

As an obvious consequence of the preceding results we have:

Proposition 3.40 *The formulas*

$$
f(x) = \sum_{n=-\infty}^{+\infty} c_n \, e^{in\omega x} \quad \text{and} \quad c_n = \frac{1}{T} \int_T e^{-in\omega x} f(x) \, dx
$$

establish a one-to-one correspondence between periodic distributions f (with period $T = 2\pi/\omega$) and tempered sequences $(c_n)_{n \in \mathbb{Z}}$.

From the continuity of derivation and translation, and from proposition 3.35, it also follows that, if

$$
f(x) = \sum_{n=-\infty}^{+\infty} c_n \, e^{in\omega x} \quad \text{and} \quad g(x) = \sum_{n=-\infty}^{+\infty} d_n \, e^{in\omega x}
$$

then

$$
\mathbf{D}f(x) = \sum_{n=-\infty}^{+\infty} (in\omega c_n) e^{in\omega x},
$$

$$
f(x - \tau) = \sum_{n=-\infty}^{+\infty} \left(e^{-in\omega \tau} c_n \right) e^{in\omega x}
$$

and

$$
f(x) \otimes g(x) = \sum_{n=-\infty}^{+\infty} (T c_n d_n) \, e^{in\omega x}.
$$

In particular the Fourier coefficients of order n of the distribution $\tilde{\delta}_T^{(k)}$ and $\tilde{\delta}_T(x - k\tau)$ are, respectively,

$$
\frac{(in\omega)^k}{T} \quad \text{and} \quad \frac{1}{T} e^{-in\omega k\tau}.
$$

An immediate consequence of proposition 3.40 which is worthwhile to remark is the *uniqueness of the Fourier coefficients* of a periodic distribution; if we have

$$
\sum_{n=-\infty}^{+\infty} c_n \, e^{in\omega x} = \sum_{n=-\infty}^{+\infty} c_n^* \, e^{in\omega x}
$$

then, denoting by $f(x)$ the sum of either of those series, we have

$$c_n \;=\; \frac{1}{T} \int_T e^{-in\omega x} f(x)\, dx \;=\; c_n^*$$

for every $n \in \mathbb{Z}$.

In some situations instead of the exponential form of the Fourier series that we have used so far, it is preferable to use the trigonometric form; if in (3.23) the terms corresponding to symmetric values of n are associated and if we set $\alpha_0 = c_0$ and, for each $n > 0$,

$$\alpha_n = c_n + c_{-n} \qquad \beta_n = i(c_n - c_{-n})$$

then it will be seen immediately that any periodic distribution may also be represented by a series of sines and cosines:

$$\alpha_0 + \alpha_1 \cos x + \beta_1 \sin x + \alpha_2 \cos 2x + \beta_2 \sin 2x + \ldots$$

The use of the trigonometric form is particularly advantageous if the distribution f is even or odd; in the first case $(\beta_n = 0)$ we get a cosine series, and in the second case $(\alpha_n = 0)$, a sine series.

In classical Analysis the theory of Fourier Series involves some very delicate points which we do not consider in this resumé of the theory in a distributional sense. On the other hand we can draw attention to the simplicity and elegance of the theory in this new setting. Among the results which might be included in a classical analysis context we may quote, in particular, the following proposition:

Proposition 3.41 *If F is a periodic function of class C^2 then its Fourier series converges to F, absolutely and uniformly on $\mathbb{R}$.*

Proof: If we denote by c_n and d_n respectively the Fourier coefficients of order n of the functions F and F'' then, as shown above, $d_n = (in\omega)^2 c_n$. Hence denoting by M the maximum of the function F'', we get for $x \in \mathbb{R}$ and $n \neq 0$,

$$\left| c_n e^{in\omega x} \right| \;=\; |c_n| \;=\; \frac{1}{n^2 \omega^2} |d_n| \;=\; \frac{1}{n^2 \omega^2} \frac{1}{T} \left| \int_0^T e^{-in\omega x} F''(x)\, dx \right|$$

$$\leq \; \frac{1}{n^2 \omega^2} \frac{1}{T} \int_0^T |F''(x)|\, dx \; \leq \; \frac{M}{n^2 \omega^2}$$

which proves the absolute and uniform convergence of the Fourier series of the function F. Since the convergence is uniform, the sum of this series will be a function $G \in C(\mathbb{R})$. However, as the uniform convergence implies convergence in the distributional sense, this series will also converges to G in $C_\infty(\mathbb{R})$, and therefore, by proposition 3.38 and by the uniqueness of the limit we must have $G = F$. $\square$

We have seen above that, given a fixed number $T > 0$, there exists a one-to-one correspondence between periodic distributions which admit T as period and the tempered sequences of complex numbers. Under this correspondence, the convolution of distributions is transformed (apart from a factor T) into the product of the corresponding sequences; if we have

$$f(x) \;=\; \sum_{n=-\infty}^{+\infty} c_n e^{in\omega x} \quad \text{and} \quad g(x) \;=\; \sum_{n=-\infty}^{+\infty} d_n e^{in\omega x} \tag{3.26}$$

then the following formula

$$f(x) \otimes g(x) \;=\; \sum_{n=-\infty}^{+\infty} T c_n d_n e^{in\omega x}$$

holds. We can take advantage of this result to simplify the solution of convolution equations in the algebra Π_∞. As we know, such an equation

$$f \otimes u \;=\; g \tag{3.27}$$

(where f, g are given periodic distributions and u is the unknown) has a unique solution in Π_∞ if and only if f is invertible in that algebra, that is, if there exists a distribution $w \in \Pi_\infty$ such that

$$f \otimes w \;=\; \tilde{\delta}_T. \tag{3.28}$$

The distribution w is called the *elementary solution* of the equation (3.27). If such a solution does exist and if $(s_n)_{n\in\mathbb{Z}}$ denotes the corresponding sequence of its Fourier coefficients then it follows from (3.28) that we must have

$$c_n s_n \;=\; \frac{1}{T^2} \quad (n \in \mathbb{Z}).$$

Hence an elementary solution exists only if

i) for each $n \in \mathbb{Z}$, we have $c_n \neq 0$;

ii) once i) is satisfied, the sequence $(1/c_n)_{n\in\mathbb{Z}}$ is tempered.

Under these hypothesis we will have

$$w(x) \;=\; \sum_{n=-\infty}^{+\infty} \frac{1}{T^2} \frac{1}{c_n} e^{in\omega x}$$

and the unique solution of (3.27) will be

$$u(x) \;=\; w(x) \otimes g(x) \;=\; \sum_{n=-\infty}^{+\infty} \frac{1}{T} \frac{d_n}{c_n} e^{in\omega x}.$$

If some of the coefficients c_n are zero, (more specifically if for a nonempty subset $\mathbb{I}$ of $\mathbb{Z}$ we have $c_n = 0$ for $n \in \mathbb{I}$ and $c_n \neq 0$ for $n \in \mathbb{Z}\backslash\mathbb{I}$), then the equation (3.27) will be solvable only if we have both $d_n = 0$ for $n \in \mathbb{I}$ and the existence of a tempered sequence

$(s_n)_{n \in \mathbb{Z}}$ such that $c_n s_n = d_n$ for each $n \in \mathbb{Z} \setminus \mathbf{I}$. If these conditions are satisfied then it is easy to see that the equation will actually have infinitely many solutions.

As an example consider the differential equation

$$a_0 \mathbf{D}^p u + a_1 \mathbf{D}^{p-1} u + \ldots + a_{p-1} \mathbf{D} u + a_p u \; = \; g, \quad (a_0 \neq 0) \tag{3.29}$$

which may be represented in the form $f \otimes u = g$, by setting

$$f \; = \; a_0 \tilde{\delta}_T^{(p)} + a_1 \tilde{\delta}_T^{(p-1)} + \ldots + a_{p-1} \tilde{\delta}_T' + a_p \tilde{\delta}_T.$$

The Fourier coefficient of order n of the distribution f is

$$c_n \; = \; \frac{1}{T} \left[a_0 (in\omega)^p + a_1 (in\omega)^{p-1} + \ldots + a_p \right].$$

Hence, the equation (3.29) will have a unique solution in the algebra Π_∞ for any given periodic distribution g on the right-hand side if and only if none of the numbers $in\omega$ $(n \in \mathbb{Z})$ is a root of the polynomial

$$P(z) \; = \; a_0 z^p + a_1 z^{p-1} + \ldots + a_{p-1} z + a_p. \tag{3.30}$$

In fact in such a situation we have $c_n \neq 0$ for each $n \in \mathbb{Z}$ and the sequence with general term

$$\frac{1}{c_n} \; = \; \frac{T}{P(in\omega)}$$

converges to zero and is therefore tempered.

Similarly, if τ is a positive real number, the difference equation

$$a_0 u(x + p\tau) + a_1 u(x + p\tau - \tau) + \ldots + a_p u(x) \; = \; g(x), \quad (a_0 \neq 0) \tag{3.31}$$

is equivalent, (in the algebra Π_∞^0), to the convolution equation $f \otimes u = g$ where we now have

$$f(x) \; = \; a_0 \tilde{\delta}_T(x + p\tau) + a_1 \tilde{\delta}_T(x + p\tau - \tau) + \ldots + a_p \tilde{\delta}_T(x).$$

Since the Fourier coefficient of order n of f is

$$c_n \; = \; \frac{1}{T} \left(a_0 e^{in\omega p\tau} + a_1 e^{in\omega(p-1)\tau} + \ldots + a_n \right),$$

a sufficient condition for the existence and uniqueness of the solution for the equation (3.31) for any right-hand side is that the polynomial (3.30) has no root with modulus equal to 1. In fact, under such hypothesis, the minimum of $|P(z)|$ for $|z| = 1$ will be a positive number M and, if

$$\frac{1}{c_n} \; = \; \frac{T}{P(e^{in\omega\tau})}$$

then we will have

$$\left| \frac{1}{c_n} \right| \; = \; \frac{T}{|P(e^{in\omega\tau})|} \; \leq \; \frac{T}{M}$$

and the sequence $(1/c_n)_{n \in \mathbb{Z}}$ will be bounded and therefore tempered.

3.4 Global Distributions (Schwartz Distributions). Support of a Distribution. The Piecewise Joining Principle.

As remarked in the Introduction, the process adopted by Schwartz to define the concept of distribution is different from the axiomatic method, due to J. Sebastião e Silva, which we have been using here. Apart from the difference in the method adopted here the very concept of distribution itself is not equivalent to the general concept of distribution in the sense of Schwartz. What has been considered as a distribution in the present text corresponds to what Schwartz describes as a *distribution of finite order*. In what follows the notion of distribution in the sense of Schwartz will be introduced under the designation of **global distribution**.

It should be stated at once that for the majority of applications the concept of global distribution may be dispensed with. Although it arises naturally in the context of the development of the theory in terms of functionals it has a more artificial aspect when, as here, a more direct approach to distributions is adopted.

In order to deal with the concept of global distribution (with domain in an arbitrary nonempty open set $A \subset \mathbb{R}$) it is useful to adopt the conventional notation $k(A)$ for the family of all (non-degenerate) compact intervals contained in A. We may then associate with each function $F \in \mathcal{C}(A)$, (that is, with each function defined and continuous on A) a certain family of continuous functions indexed by the members of $k(A)$; specifically, with each $F \in \mathcal{C}(A)$ we associate the family $\{F_K\}_{K \in k(A)}$ where, for each K, F_K denotes the restriction of F to the interval K.

On the other hand, if we are given an arbitrary family $\{F_K\}_{K \in k(A)}$ such that F_K belongs to $\mathcal{C}(K)$ for each K, it is clear that there will not always exist a function $F \in \mathcal{C}(A)$ which satisfies the condition that $F_K = F_{|K}$, for each $K \in k(A)$; however, it is easy to see that for such a function F to exist it is necessary and sufficient that the family $\{F_K\}_{K \in k(A)}$ is *compatible*, in the following sense: whenever we have $K, L \in k(A)$ and $K \subset L$ then F_K must coincide with the restriction to K of the function F_L.

Suppose now that A is an *open interval*. Under this condition, if f is a distribution defined on A it will similarly be possible to associate with f a family $\{f_K\}_{K \in k(A)}$ of distributions which is again *compatible*; that is, $f_K = \rho_{KL}(f_L)$ if $K \subset L$ and K, L are in $k(A)$. However it is also easy to see that, in this case, there will exist compatible families of distributions which do not correspond, in the given sense, to any distribution defined on the interval A itself. If, for example, $(x_n)_{n \in \mathbb{N}}$ is a sequence of points in A that has as its limit an end point (finite or infinite) of that interval, we may define a compatible family of distributions $\{f_K\}_{K \in k(A)}$ associating with each $K \in k(A)$ the distribution:[12]

$$f_K(x) = \sum_{x_n \in K} \delta^{(n)}(x - x_n).$$

---

[12]As the notation suggests, for each $K \in k(A)$ the distribution f_K is the sum of all terms of the form $\delta^{(n)}(\hat{x} - x_n)$ corresponding to values of n such that $x_n \in K$.

Clearly this family will contain distributions of arbitrarily large degree and therefore there cannot exist any distribution $f \in \mathcal{C}_{\infty}(A)$ such that $f_{|K} = f_K$, for each $K \in k(A)$.

Now we will start from these ideas to generalise the concept of distribution. This generalization will be made simultaneously in two senses: on the one hand (in case A is an open interval) we will introduce a new type of distributions said to be of *infinite order* or *degree*; on the other hand, we will consider distributions whose domains are not necessarily intervals but instead may be arbitrary (nonempty) open sets. As already remarked, to avoid confusion with the notion originally introduced in section 1.3, we will use the term *global distribution* to denote the new sense of the word. If A is a nonempty open set in $\mathbb{R}$ and $k(A)$ denotes the set of all (non-degenerate) compact intervals contained in A, we will give the name *global distribution* to any compatible family of distributions indexed by $k(A)$; more precisely, to any family $\{f_K\}_{K \in k(A)}$ which satisfies the following conditions:

i) $f_K \in \mathcal{C}_{\infty}(K)$, for each $K \in k(A)$;

ii) whenever $K, L \in k(A)$ and $K \subset L$ we will have $\rho_{KL}(f_L) = f_K$.

The set of all global distributions defined on A will be denoted by $\mathcal{C}_{\pi}(A)$.

Whenever there is no danger of confusion, instead of the expression $\{f_K\}_{K \in k(A)}$ we will write simply $\{f_K\}$. For each $K \in k(A)$, the distribution f_K is called a *component* of f, more precisely, the Kth-*component* of the global distribution f.

The sum of global distributions, and the product by scalars, can be defined componentwise in $\mathcal{C}_{\pi}(A)$: if $f = \{f_K\}$, $g = \{g_K\}$ and $\alpha \in \mathbf{C}$ we set by definition

$$f + g = \{f_K + g_K\}, \qquad \alpha f = \{\alpha f_K\}$$

(it is obvious that if $\{f_K\}$ and $\{g_K\}$ are compatible families the same will be true for $\{f_K + g_K\}$ and $\{\alpha f_K\}$).

With these operations $\mathcal{C}_{\pi}(A)$ is a complex linear space.

Similarly, if φ is a function of class $\mathcal{C}^{\infty}$ defined on the set A, the product of φ with the global distribution $\{f_K\}$ can be defined by means of the formula

$$\varphi f = \{\varphi f_K\}.$$

The concept of derivative is also defined componentwise. If $f = \{f_K\}$ then the derivative of f, $\mathbf{D}f$, is the global distribution $\{\mathbf{D}f_K\}$; defined in this way, the derivative is still a linear map from $\mathcal{C}_{\pi}(A)$ into itself.

If, as described at the beginning, we associate with each function $F \in \mathcal{C}(A)$ the global distribution $\{F_K\}$, where F_K denotes the restriction of F to the interval K, we will obtain a mapping from $\mathcal{C}(A)$ into $\mathcal{C}_{\pi}(A)$ which is injective (since for two distinct functions $F, G \in \mathcal{C}(A)$ we cannot have $F_K = G_K$ for every $K \in k(A)$). We may therefore identify $\mathcal{C}(A)$ with a linear subspace of $\mathcal{C}_{\pi}(A)$. Any function F, which is continuous on A, now has derivatives of all orders, which will be global distributions defined on A.

The global distributions representable in the form $f = \mathbf{D}^p F$, with $p \in \mathbb{N}$ and $F \in \mathcal{C}(A)$, are said to be *global distributions of finite degree* (or *of finite order*); we may also call them simply *distributions* defined on A. These also constitute a linear subspace of $\mathcal{C}_\pi(A)$ which will be denoted by $\mathcal{C}_\infty(A)$. In the particular case when A is an open interval, it is easy to see that $\mathcal{C}_\infty(A)$ coincides with the space of the distributions defined on A in the sense considered in the preceding paragraphs. This ensures the coherence of the notations adopted. Similarly we define the spaces $\mathcal{C}_p(A)$ when $A \subset \mathbb{R}$ is an arbitrary open set and $p \in \mathbb{N}$.

Now let A and B be two nonempty open subsets of $\mathbb{R}$, with $B \subset A$, and let $f \in \mathcal{C}_\pi(A)$, $f = \{f_K\}_{K \in k(A)}$. The element of $\mathcal{C}_\pi(B)$ defined by

$$f_{|B} = \{f_K\}_{K \in k(B)}$$

is called the restriction of f to B, and is denoted by $f_{|B}$ or $\rho_{BA}(f)$. If A and B are two open sets of $\mathbb{R}$ with nonempty intersection and D is a nonempty open set contained in $A \cap B$, we say that $f \in \mathcal{C}_\pi(A)$ and $g \in \mathcal{C}_\pi(B)$ are equal on D if $f_{|D} = g_{|D}$. We then write $f = g$ on D. In particular we will say that the global distribution $f \in \mathcal{C}_\pi(A)$ is null on D if f is equal, in D, to the null global distribution (identified with the null function).

We will now introduce the following theorem

> **Theorem 3.42** *Let B be an open set of $\mathbb{R}$ and f a global distribution defined on B; also let $\{A_i\}_{i \in J}$ be a (finite or infinite) family of open subsets of B. If f is null in each one of the sets A_i then f is null on its union $A = \cup_{i \in J} A_i$.*

The proof will depend on the following lemma which we will prove first.

> **Lemma 3.43** *Let $K = [a, b]$ be a compact interval of $\mathbb{R}$ and $\{A_i\}_{i \in J}$ a family of open subsets of $\mathbb{R}$ which covers K (that is, such that K is contained in $\cup_{i \in J} A_i$). Then there exists a positive number ε such that each subinterval of K with length $\leq \varepsilon$ is contained in some A_i; more precisely there exists $\varepsilon > 0$ such that, whenever x and y are two points of K such that $x \leq y \leq x + \varepsilon$ we will have $[x, y] \subset A_i$ for some $i \in J$.*

Proof: Suppose not. Then for each $n \in \mathbb{N}_1$ there would exist $x_n, y_n \in K$ such that $x_n < y_n < x_n + \frac{1}{n}$, with $[x_n, y_n] \not\subset A_i$, for each $i \in J$ and each $n \in \mathbb{N}$.

By the Bolzano-Weierstrass Theorem the bounded sequence $(x_n)_{n \in \mathbb{N}}$ would admit a subsequence $(x_{p_n})_{n \in \mathbb{N}}$ which would converge to a certain point α, to which the sequence $(y_{p_n})_{n \in \mathbb{N}}$ would also converge. Since $x_{p_n} \in K$ and K is compact it would follow that $\alpha \in K \subset \cup_{i \in J} A_i$.

Thus there would exist $j \in J$ such that $\alpha \in A_j$. Since A_j is open and since $(x_{p_n})_{n \in \mathbb{N}}$ and $(y_{p_n})_{n \in \mathbb{N}}$ both converge to α, it follows that for sufficiently large values of n the intervals $[x_{p_n}, y_{p_n}]$ should be contained in A_j contrary to the hypothesis that $[x_n, y_n] \not\subset A_i$ for each $i \in J$ and each $n \in \mathbb{N}$. $\qquad\square$

Proof: (Theorem 3.42) To prove that the global distribution $f = \{f_K\}$ which, by hypothesis, is null on each of the sets A_i, is itself null on $A = \cup_{i \in J} A_i$, it is enough to show that, whenever K is a non-degenerate compact interval contained in A, we have $f_K = 0$.

Accordingly, let $K \in k(A)$. By lemma 3.43 there will exist $\varepsilon > 0$ such that each subinterval of K with length $\leq \varepsilon$ will be contained in some set A_i. It follows that, if we fix a partition in $K = [a, b]$ determined by the points

$$a = x_0 < x_1 < \cdots \cdots < x_{n-1} < x_n = b$$

chosen so that $x_i - x_{i-1} < \frac{\varepsilon}{2}$ for $i = 1, 2, \ldots, n$, each one of the intervals

$$K_1 = [x_0, x_2], \ K_2 = [x_1, x_3], \ \ldots, \ K_{n-1} = [x_{n-2}, x_n]$$

will be contained in some A_i and therefore the restriction of f_K to the interval K_s will be the null distribution $(s = 1, \ldots, n - 1)$.

Hence, representing f_K in the form $f_K = \mathbf{D}^p F$, with $p \in \mathbb{N}$ and $F \in \mathcal{C}(K)$, the restriction of F to each one of the intervals K_s must be a polynomial P_s of degree $< p$. However, since for $s = 1, \ldots, n - 2$ the polynomials P_s and P_{s+1} must coincide in $K_s \cap K_{s+1} = [x_s, x_{s+1}]$, it can be seen immediately that the function F itself must be a polynomial of degree $< p$, that is, that $f_K = 0$. $\qquad\square$

We will now introduce a notion of great importance, namely the notion of support of a global distribution.

First recall that, if F is a continuous function on $\mathbb{R}$, it is usual to call the support of F the complement (with respect to $\mathbb{R}$) of the "maximal" open set on which F is null (that is, the union of all open sets on which F is null). As the complement of an open set, the support of F is a closed set; and it is easily seen that it coincides with the closure of the set of points x where $F(x) \neq 0$. For example, the support of the function $F(x) = \sin x + |\sin x|$ is the set of points x such that $\sin x \geq 0$.

Now if f is a global distribution defined on $\mathbb{R}$ we will similarly define the *support* of f to be the complement of the maximal open set on which $f = 0$, and denote it by **supp**f. (The existence of this "maximal" set on which f is null has been proved in the theorem 3.42). Whenever $x \in \mathbf{supp}f$ and V is any open set such that $x \in V$, the condition $f = 0$ in V will be false; on the other hand, for each $y \in \mathbb{R}\backslash\mathbf{supp}f$ there will exist an open set W such that $y \in W$ and $f = 0$ on W. For example, for the global distribution (of finite degree) defined on $\mathbb{R}$ by the formula

$$f(x) = e^{x-|x|} - 1 + \delta(x) - \delta'(x - 1)$$

we have $\mathbf{supp}f =]-\infty, 0] \cup \{1\}$.

More generally, if f is a global distribution defined on an open set $A \subset \mathbb{R}$ then the *support* of f, again denoted by $\mathbf{supp}f$, will be the complement in A of the union of all open sets on which f is null. The support of f is always a closed set with respect to A (that is, if $(x_n)_{n \in \mathbb{N}}$ is a sequence of points from $\mathbf{supp}f$, converging to a point $\alpha \in A$ then we always have $\alpha \in \mathbf{supp}f$).

It is easy to see that if the global distribution $f \in \mathcal{C}_\pi(A)$ can be identified with a function continuous on A, then the support of f as a global distribution coincides with its support as a continuous function (that is, with the complement in A of the maximal open set on which the function f is null). Also, if A is an open interval and f a finite degree distribution, $f \in \mathcal{C}_\infty(A)$, the support of f as a global distribution coincides with the complement of the union of all open intervals on which f is null.

Another very important result is the following theorem of Schwartz:

Theorem 3.44 (Piecewise Joining Principle) *Suppose that $\{A_i\}_{i \in J}$ is a family of nonempty open subsets of $\mathbb{R}$ and suppose that with each $i \in J$ there is associated a global distribution f_i such that the following conditions hold:*

(1) *$f_i \in \mathcal{C}_\pi(A_i)$, for each $i \in J$;*

(2) *whenever $i, j \in J$ and $A_i \cap A_j \neq \emptyset$ we have $f_i = f_j$ on $A_i \cap A_j$.*

Under these conditions, if $A = \cup_{i \in J} A_i$, then there exists one and only one global distribution $g \in \mathcal{C}_\pi(A)$ such that $g_{|A_i} = f_i$, for each $i \in J$.

Proof: The uniqueness is an immediate consequence of the theorem 3.42: the difference between two global distributions satisfying the stated conditions is a distribution which is null in each one of the sets A_i and therefore is the null distribution on A.

To prove the existence we must show that it is possible, under the hypothesis, to associate with each interval $K \in k(A)$ a distribution g_K, such that the family $\{g_K\}_{K \in k(A)}$ is compatible and that $f_i = \{g_K\}_{K \in k(A_i)}$ for each $i \in J$.

To begin with, given the interval $K = [a, b] \in k(A)$, we remark that, as in the proof of Theorem 3.42, the lemma 3.43 ensures that it is possible to obtain a partition of K, determined by the points

$$a = x_0 < x_1 < \cdots \cdots < x_n = b,$$

such that each one of the intervals

$$K_1 = [x_0, x_2], \ K_2 = [x_1, x_3], \ \ldots\ldots, \ K_{n-1} = [x_{n-2}, x_n]$$

is contained in one open set A_i.

Denoting in general by A_{i_s} an open set of the family in question which contains the interval K_s and by g_s the component of the distribution f_{i_s} corresponding to the interval K_s then we will have, in accordance with condition **(2)** of the statement of the theorem, $g_s = g_{s+1}$ in $[x_s, x_{s+1}]$, since this interval is contained in $A_{i_s} \cap A_{i_{s+1}}$.

Hence, representing each g_s in the form $g_s = \mathbf{D}^p G_s$, with $G_s \in \mathcal{C}(K_s)$, there must exist for each s a polynomial P_s, of degree $< p$, such that

$$G_s(x) - G_{s+1}(x) = P_s(x), \quad (x \in [x_s, x_{s+1}]).$$

If we set

$$G(x) = \begin{cases} G_1(x) & \text{if } \ x \in [x_0, x_2] \\ G_2(x) + P_1(x) & \text{if } \ x \in \,]x_2, x_3] \\ G_3(x) + P_1(x) + P_2(x) & \text{if } \ x \in \,]x_3, x_4] \\ \cdots & \\ G_{n-1}(x) + P_1(x) + \cdots + P_{n-2}(x) & \text{if } \ x \in \,]x_{n-1}, x_n] \end{cases}$$

then it is easy to see that a function $G \in \mathcal{C}(K)$ is defined and also that the distribution $g_K = \mathbf{D}^p G$ is a common extension of the distributions g_s:

$$g_{K|K_s} = g_s, \quad s = 1, \ldots, n - 1.$$

Now the following question naturally arises: if, instead of the partition of K referred to above, we had considered another

$$a = y_0 < y_1 < \cdots\cdots < y_m = b$$

(so that each one of the intervals $[y_0, y_2], [y_1, y_3], \ldots, [y_{m-2}, y_m]$ was also contained in some set A_i) would we have obtained the same distribution g_K by carrying out the process described?

It is easy to prove that the answer is in the affirmative if we first assume that the second partition is finer than the first (that is, that it contains all the points of the first, eventually with additional ones). Note that then the case of two arbitrary partitions is easily reduced to this one by comparing each one of the two partitions with their common refinement (that is, with the partition containing both all the points of the first and all the points of the second).

The process just described allow us to associate with each interval $K \in k(A)$ a distribution $g_K \in \mathcal{C}_\infty(K)$, determined without any ambiguity.

There is also no difficulty in seeing that, if $K, L \in k(A)$ with $K \subset L$, then we necessarily have $g_K = \rho_{KL}(g_L)$ (we may, for example, use a partition of the interval L to determine g_L which is an "extension" of the partition of K used to determine g_K). It follows then that the family $\{g_K\}$ is, in fact, a global distribution (defined on A).

Finally, if $i \in J$ and $K \in k(A_i)$, the distribution g_K may be determined by considering the partition of K constituted by only one interval, which allows us to see immediately that g_K coincides with the Kth component of the distribution f_i; hence, we have $g_{|A_i} = f_i$, for each $i \in J$ and this ends the proof. $\qquad\square$

As was clearly indicated when the concept of global distribution was first introduced the piecewise joining principle is not valid in the finite degree distributions setting: if we admit that each one of the global distributions f_i referred to in the theorem 3.44 has finite degree, p_i, we cannot conclude that g necessarily has finite degree. However, the following theorem holds:

> **Theorem 3.45** *Suppose that, in addition to the conditions stated in the hypothesis of the theorem 3.44, each one of the distributions f_i has finite degree, p_i; moreover suppose also that there exists a number p such that $p_i \leq p$ for each $i \in J$. Then the distribution g has finite degree not greater than p.*

Proof: Consider first the case when A is an open interval. Let $K_1, K_2, \ldots, K_n, \ldots$ be a sequence[13] of nondegenerate compact intervals, such that

$$K_1 \subset K_2 \subset \cdots\cdots \subset K_n \subset \cdots\cdots \quad \text{and} \quad A = \cup_{n \in \mathbb{N}_1} K_n.$$

For each interval K_n construct a distribution g_{K_n}, or more simply g_n, as in the initial part of the proof of theorem 3.44. Clearly each one of the distributions g_n will have a degree $\leq p$

---

[13]To recall how can we obtain such a sequence, see again the proof of theorem 1.11.

(cf. proof of theorem 3.44) and may therefore be represented in the form $g_n = \mathbf{D}^p G_n$, with $G_n \in \mathcal{C}(K_n)$.

Since, for every n, we must have $g_n = g_1$ in K_1, for every n there must exist a polynomial P_n such that

$$G_n(x) - G_1(x) \; = \; P_n(x), \qquad (x \in K_1).$$

It is then easy to see that a function $G \in \mathcal{C}(A)$ may be defined by setting for each n and each $x \in K_n$,

$$G(x) \; = \; G_n(x) - P_n(x)$$

and by appealing in particular to theorem 1.11, it is also easy to see that the distribution $\mathbf{D}^p G$ necessarily coincides with g, which is therefore of finite degree.

Now let $A \subset \mathbb{R}$ be an arbitrary nonempty open set; it is then easy to recognize that A is the union of a family of open intervals, $\mathbf{I}_j$, which are mutually disjoint.[14]

Representing the restriction of f to $\mathbf{I}_j$ in the form $\mathbf{D}^p F_j$ with $F_j \in \mathcal{C}(\mathbf{I}_j)$ (which is always possible, according to the first part of this proof), we will obtain obviously $f = \mathbf{D}^p F$, where F is the function continuous in A defined by $F(x) = F_j(x)$, for each j and each $x \in \mathbf{I}_j$. $\square$

We observe now that any global distribution defined on $\mathbb{R}$ has finite degree in any bounded interval $]a, b[$: it is enough to note that the component of f relative to the interval $[a, b]$ is of finite degree; more generally, if A is an open subset of $\mathbb{R}$ and $f \in \mathcal{C}_\pi(A)$, f is of finite degree in any open interval whose closure is contained in the set A. From this fact and the theorem 3.45, it follows easily that

> **Corollary 3.46** *Let f be a global distribution in $\mathbb{R}$ and suppose that there exist $\alpha, \beta \in \mathbb{R}$ such that f is of finite degree both on $] - \infty, \alpha[$ and on $]\beta, +\infty[$. Then f is a distribution of finite degree, that is $f \in \mathcal{C}_\infty(\mathbb{R})$.*

In particular any global distribution on $\mathbb{R}$ with compact support will be a finite order distribution (which allows us to see that the unique global distributions with support contained in a singleton set, $\{a\}$, are the linear combinations of derivatives of the distribution $\delta(x - a)$).

It is convenient to observe also that the concepts of value and limit of a distribution at a point may be generalised to global distributions in an obvious way. We will only give some examples here and, for the sake of simplicity, restrict attention to global distributions defined on $\mathbb{R}$. (There is no real difficulty in the extension to distributions defined on an arbitrary open set.) Naturally, we will say that a global distribution $f \in \mathcal{C}_\pi(\mathbb{R})$ is continuous at the point $a \in \mathbb{R}$ and that $f(a) = \alpha$, if and only if there exists a compact interval K such that a is in the interior of K, the component of f with respect to K, f_K, is continuous at the point a (in the sense considered in section 2.1) and $f_K(a) = \alpha$.

---

[14]If a is any point in A, the interval I_j which contains a (called the connected component determined by the point a) has as lower bound the supremum (finite or $-\infty$) of the set of all points $x < a$ such that $x \notin A$, and as upper bound the infimum (finite or $+\infty$) of the set of points $x > a$ such that $x \notin A$. It is easy to see that the connected components determined by two points $a, b \in A$ either coincide or are disjoint. Taking each connected component just once, we obtain A as a union of open intervals mutually disjoint.

Similarly, we will say that $\lim_{x\to+\infty} f(x) = \alpha$ if there exists an interval $\mathbf{I}$, not bounded above, such that the restriction of f to $\mathbf{I}$ is a distribution of finite degree, with limit α at the point $+\infty$ in the sense of section 2.2.

The integrability (of global distributions) in $\mathbb{R}$ will be defined only for distributions of finite degree, therefore coinciding with the concept introduced in section 2.5; a particularly important case of distributions which are integrable on $\mathbb{R}$ occurs in the case of the elements of $\mathcal{C}_\infty(\mathbb{R})$ which have compact support, as already seen in proposition 2.44.

We now take the opportunity to give an idea of the definition of distribution adopted by L. Schwartz. For this purpose it is convenient to introduce the space $\mathcal{D}(\mathbb{R})$ (or, simply $\mathcal{D}$) comprising all functions defined on $\mathbb{R}$, which are infinitely differentiable and which have compact support. It is not perhaps evident that functions of this type, other than the null function, really exist; however an example of such a function is the function θ defined by:

$$\theta(x) = \begin{cases} \exp\left(-\frac{1}{1-x^2}\right) & \text{if } |x| < 1 \\ 0 & \text{if } |x| \geq 1. \end{cases}$$

The function θ is clearly indefinitely differentiable at any point other than $+1$ or -1; at each of these two points, however, it is also not difficult to see that θ has derivatives of all orders equal to 0. The support of θ is the interval $[-1, +1]$.

With the usual operations of addition of functions and multiplication of a function by a complex number, the set $\mathcal{D}$ is clearly a complex linear space. We will introduce in $\mathcal{D}$ a notion of convergence of sequences by saying that a sequence $(\varphi_n)_{n\in\mathbb{N}}$, with terms in $\mathcal{D}$, converges in this space to the function $\varphi \in \mathcal{D}$ if and only if the two following conditions hold:

i) for each $p \in \mathbb{N}$, the sequence $(\varphi_n^{(p)})_{n\in\mathbb{N}}$ converges uniformly to the function $\varphi^{(p)}$ (that is, the sequence $(\varphi_n)_{n\in\mathbb{N}}$ converges to φ in $\mathcal{C}^\infty(\mathbb{R})$);

ii) there exists a compact $K \subset \mathbb{R}$ such that $\mathbf{supp}\varphi_n \subset K$, for every integer n in $\mathbb{N}$.

Thus, if θ is the function considered in the example above, the sequence $(\frac{1}{n}\theta(x))_{n\in\mathbb{N}}$ converges in $\mathcal{D}$ to the null function, but the sequence $(\frac{1}{n}\theta(x-n))_{n\in\mathbb{N}}$ does not converge in $\mathcal{D}$ (but only in $\mathcal{C}^\infty$).

Now let $f = \{f_K\}$ be an arbitrary global distribution defined on $\mathbb{R}$. For each function $\varphi \in \mathcal{D}$ the product $\varphi f = \{\varphi f_K\}$ will be a distribution of compact support, therefore integrable on $\mathbb{R}$; and if the compact interval $K = [a, b]$ contains the support of φ, we will have, supposing $f_K = \mathbf{D}^p F$:

$$\int_{\mathbf{R}} \varphi f = \int_a^b \varphi \mathbf{D}^p F = -\int_a^b \varphi' \mathbf{D}^{p-1} F = \cdots = (-1)^p \int_a^b \varphi^{(p)} F,$$

as is easily verified by using integration by parts and taking into account that φ and all its derivatives are zero at the points a and b.

Hence, for a given global distribution $f = \{f_K\} \in \mathcal{C}_\pi(\mathbb{R})$, we may associate with each function $\varphi \in \mathcal{D}$ a complex number $\Theta(\varphi)$ through the formula

$$\Theta(\varphi) = \int_{\mathbb{R}} \varphi f.$$

If $\varphi, \psi \in \mathcal{D}$ and $\alpha, \beta \in \mathbf{C}$, we will have

$$\begin{aligned}
\Theta(\alpha\varphi + \beta\psi) &= \int_{\mathbb{R}} (\alpha\varphi + \beta\psi) f \\
&= \alpha \int_{\mathbb{R}} \varphi f + \beta \int_{\mathbb{R}} \psi f = \alpha\Theta(\varphi) + \beta\Theta(\psi).
\end{aligned} \tag{3.32}$$

Moreover, suppose that $(\varphi_n)_{n\in\mathbb{N}}$ is a sequence in $\mathcal{D}$ converging to the function φ, and that $K = [a, b]$ is a compact interval containing the supports of all the φ_n (and therefore the supports of the function φ and all its derivatives). Then if $f_K = \mathbf{D}^p F$ we have

$$\begin{aligned}
|\Theta(\varphi_n) - \Theta(\varphi)| &= \left| \int_{\mathbb{R}} (\varphi_n - \varphi) f \right| = \left| \int_a^b (\varphi_n - \varphi) \mathbf{D}^p F \right| \\
&= \left| \int_a^b \left(\varphi_n^{(p)} - \varphi^{(p)} \right) F \right| \leq \int_a^b \left| \varphi_n^{(p)} - \varphi^{(p)} \right| |F| \\
&\leq (b-a) \max_{x\in K} |F(x)| \sup_{x\in K} \left| \varphi_n^{(p)}(x) - \varphi^{(p)}(x) \right|,
\end{aligned}$$

from which, taking into account that $(\varphi_n^{(p)})_{n\in\mathbb{N}}$ converges to $\varphi^{(p)}$ uniformly in $\mathbb{R}$, it follows that the numerical sequence $(\Theta(\varphi_n))_{n\in\mathbb{N}}$ converges to the number $\Theta(\varphi)$.

It is usual to refer to any map from a complex linear space E into the scalar field $\mathbf{C}$ as a *functional* on E. Hence the map $\Theta : \mathcal{D} \to \mathbf{C}$ is a functional on the linear space of all infinitely differentiable functions with compact support. The relation (3.32) ensures that the functional Θ is *linear*; the fact that $\Theta(\varphi_n)$ converges to $\Theta(\varphi)$ whenever $(\varphi_n)_{n\in\mathbb{N}}$ converges to φ in $\mathcal{D}$ may be expressed by saying that the functional Θ is *continuous*.

Usually the set comprising all continuous linear functionals on $\mathcal{D}$ is called the *dual* of the space $\mathcal{D}$ and is denoted by $\mathcal{D}'$. $\mathcal{D}'$ may be given the structure of a complex linear space by defining, for $\Lambda_1, \Lambda_2 \in \mathcal{D}'$, $\alpha \in \mathbf{C}$ and $\varphi \in \mathcal{D}$:

$$(\Lambda_1 + \Lambda_2)(\varphi) = \Lambda_1(\varphi) + \Lambda_2(\varphi) \quad \text{and} \quad (\alpha\Lambda_1)(\varphi) = \alpha\Lambda_1(\varphi).$$

As we have just seen, any global distribution $f \in \mathcal{C}_\pi(\mathbb{R})$ uniquely determines a functional $\Theta \in \mathcal{D}'$; we therefore obtain a map $f \rightsquigarrow \Theta$, from $\mathcal{C}_\pi(\mathbb{R})$ into $\mathcal{D}'$ which is easily seen to be linear. It is possible to prove that this map is one to one and onto, that is, that it is an isomorphism between the linear spaces $\mathcal{C}_\pi(\mathbb{R})$ and $\mathcal{D}'$. The proof is not simple and will not be given here. We content ourselves by remarking that it is precisely the continuous linear functionals on $\mathcal{D}$, that is, the elements of $\mathcal{D}'$, which Schwartz calls distributions (on $\mathbb{R}$); the Schwartz distributions are therefore identifiable, up to an isomorphism, with the elements of the space $\mathcal{C}_\pi(\mathbb{R})$, that is, with the mathematical objects which we have called global distributions.

Appendix A
A More General Notion of Value of a Distribution at a Point

In this section[1] we introduce a concept of the value of a distribution at a point that is more general than the one considered in 2.1 and mention, without proof,[2] some properties of this concept. We could also similarly generalise the concepts of limit and one-sided limit of a distribution at a point of $\mathbb{R}$ or at the points $+\infty$ or $-\infty$. These generalisations of the definitions given in the sections 2.1 and 2.2 pave the way for an extension of the concept of integral of a distribution over an interval of $\mathbb{R}$; moreover, it is not difficult to see that all those properties of the limits and integrals already considered retain their validity in this more general setting. (In fact the same could be said about the applications of these concepts to other areas also considered in this book.)

Let $\mathbf{I}$ be an interval of $\mathbb{R}$ and let a denote an interior point of $\mathbf{I}$. We shall say that a function $G : \mathbf{I}\backslash\{a\} \to \mathbb{R}$ is **bounded near** a if and only if there exists $\varepsilon > 0$ such that the interval $\mathbf{I}_\varepsilon(a) =]a - \varepsilon, a + \varepsilon[$ is contained in $\mathbf{I}$ and the restriction of G to $\mathbf{I}_\varepsilon\backslash\{a\}$ is a bounded function. Clearly, if G is a function bounded near a, the upper and lower limits of G at the point a,

$$\overline{G(a)} = \lim_{\varepsilon\to 0^+} \sup\{ G(x) : x \in \mathbf{I}_\varepsilon(a)\backslash\{a\} \}$$

$$\text{and} \quad \underline{G(a)} = \lim_{\varepsilon\to 0^+} \inf \{ G(x) : x \in \mathbf{I}_\varepsilon(a)\backslash\{a\} \} \tag{A.1}$$

are both finite and we have $\underline{G(a)} \leq \overline{G(a)}$ (with $\underline{G(a)} = \overline{G(a)}$ if and only if the function G has a limit at the point a).

In the sequel we shall denote by $\mathcal{C}_a(\mathbf{I}, \mathbb{R})$ the set of all continuous functions $G : \mathbf{I}\backslash\{a\} \to \mathbb{R}$ that are bounded near a. For each $G \in \mathcal{C}(\mathbf{I}, \mathbb{R})$ and each $x \in \mathbf{I}\backslash\{a\}$ we shall put

$$\mu_a G(x) = \frac{1}{x - a} \int_a^x G(t)\, dt.$$

The definitions of $\underline{G(a)}$ and of $\overline{G(a)}$ immediately imply that, for each $\delta > 0$, there exists $\varepsilon > 0$ such that

$$\underline{G(a)} - \delta < G(x) < \overline{G(a)} + \delta$$

if $x \in \mathbf{I}_\varepsilon(a)\backslash\{a\}$. Then we shall also have for the same values of x:

$$\underline{G(a)} - \delta = \frac{1}{x - a} \int_a^x \left(\underline{G(a)} - \delta\right) dt$$

---

[1]This section was not included in the Portuguese version of this book.

[2]The interested reader may like to obtain the proofs by himself or else refer to [6].

$$\leq\ \mu_a G(x)\ \leq\ \frac{1}{x-a}\int_a^x \left(\overline{G(a)}+\delta\right)dt\ =\ \overline{G(a)}+\delta.$$

We conclude by saying that, if $G \in \mathcal{C}_a(\mathbf{I},\mathbb{R})$, then we have $\mu_a G \in \mathcal{C}_a(\mathbf{I},\mathbb{R})$ and

$$\underline{G(a)}\ \leq\ \underline{\mu_a G(a)}\ \leq\ \overline{\mu_a G(a)}\ \leq\ \overline{G(a)}. \tag{A.2}$$

Now, define a **real distribution** (on an interval $\mathbf{I}$) to be a distribution $g \in \mathcal{C}_\infty(\mathbf{I})$ that can be represented as $g = \mathbf{D}^n G$ with G a continuous real-valued function on $\mathbf{I}$ and $n \in \mathbb{N}$. Denote by $\mathcal{C}_\infty(\mathbf{I},\mathbb{R})$ the subset of $\mathcal{C}_\infty(\mathbf{I})$ comprising all real distributions defined on $\mathbf{I}$.

Each function $G \in \mathcal{C}_a(\mathbf{I},\mathbb{R})$, having an indefinite integral over $\mathbf{I}$, may be identified with a distribution defined on $\mathbf{I}$, more precisely with the distribution

$$\mathbf{D}\left(\int_a^x G(t)\,dt\right),$$

and it is clear that this is a real distribution. Moreover, we can see immediately that to two different functions in $\mathcal{C}_a(\mathbf{I},\mathbb{R})$ there will correspond in this way two different distributions. Hence we can write $\mathcal{C}_a(\mathbf{I},\mathbb{R}) \subset \mathcal{C}_\infty(\mathbf{I},\mathbb{R})$.

Now let us consider again the operator ∂_a defined on $\mathcal{C}_\infty$ by means of the formula

$$\partial_a f\ =\ \mathbf{D}\left[(\hat{x}-a)f(\hat{x})\right].$$

We will say that a real distribution g is **bounded near the point** a if and only if there exist $p \in \mathbb{N}$ and $G \in \mathcal{C}_a(\mathbf{I},\mathbb{R})$ such that $g = \partial_a^p G$. Denote by $\mathcal{B}_a(\mathbf{I},\mathbb{R})$ the set of all real distributions defined on $\mathbf{I}$ that are bounded near a.

For $g \in \mathcal{B}_a(\mathbf{I},\mathbb{R})$ define the **degree of g at the point** a, denoted by $\mathbf{deg}_a(g)$, to be the minimum of the set of all integers $p \geq 0$ for which there exists $G \in \mathcal{C}_a(\mathbf{I},\mathbb{R})$ such that $g = \partial_a^p G$. Since $\partial_a^p G = \partial_a^{p+1}(\mu_a G)$ then for each $n \geq \mathbf{deg}_a(g)$ there exists $G_n \in \mathcal{C}_a(\mathbf{I},\mathbb{R})$ such that $g = \partial_a^n G_n$.

Now, as is easily seen, $\mathcal{B}_a(\mathbf{I},\mathbb{R})$ is a linear subspace of the (real) linear space $\mathcal{C}_\infty(\mathbf{I},\mathbb{R})$ and the operator ∂_a, restricted to $\mathcal{B}_a(\mathbf{I},\mathbb{R})$, is an automorphism.[3] Hence, for each $g \in \mathcal{B}_a(\mathbf{I},\mathbb{R})$ and every $n \geq \mathbf{deg}_a(g)$ there exists one and only one function $G_n \in \mathcal{C}_a(\mathbf{I},\mathbb{R})$ such that $g = \partial_a^n G_n$; moreover, since for $n \geq \mathbf{deg}_a(g)$ we have $G_{n+1} = \mu_a G_n$, it follows from (A.2) that we will also have

$$\underline{G_n(a)}\ \leq\ \underline{G_{n+1}(a)}\ \leq\ \overline{G_{n+1}(a)}\ \leq\ \overline{G_n(a)}.$$

Thus we see that the sequences $\left(\underline{G_n(a)}\right)_{n\in\mathbb{N}}$ and $\left(\overline{G_n(a)}\right)_{n\in\mathbb{N}}$, the terms of which are defined for $n \geq \mathbf{deg}_a(g)$, are convergent and also that

$$\lim_{n\to\infty}\underline{G_n(a)}\ \leq\ \lim_{n\to\infty}\overline{G_n(a)}.$$

---

[3]The proof of Theorem 2.3 applies here without any major change.

The limit on the left hand side of the above inequality will be called the **lower limit** *of the distribution g at the point a* and will be denoted by $\liminf g(a)$; the limit on the right hand side of the same inequality will be called the **upper limit** *of the distribution g at the point a* and will be denoted by $\limsup g(a)$.[4]

Now, let us state some elementary properties of these limits.

Proposition A.1 *If $g, h \in \mathcal{B}_a(\mathbf{I}, \mathbb{R})$ and α and β are respectively the smallest and the greatest of the two numbers $\liminf g(a) + \limsup h(a)$ and $\liminf h(a) + \limsup g(a)$, then we have*

$$\liminf g(a) + \liminf h(a) \ \leq \ \liminf(g + h)(a) \ \leq \ \alpha \quad \text{and}$$
$$\beta \ \leq \ \limsup(g + h)(a) \ \leq \ \limsup g(a) + \limsup h(a).$$

Proposition A.2 *If φ is a real C^∞-function defined on $\mathbf{I}$ and $g \in \mathcal{B}_a(\mathbf{I}, \mathbb{R})$, then $\varphi g \in \mathcal{B}_a(\mathbf{I}, \mathbb{R})$ and we have*

$$\liminf(\varphi g)(a) \ = \ \varphi(a) \liminf g(a),$$
$$\limsup(\varphi g)(a) \ = \ \varphi(a) \limsup g(a)$$

if $\varphi(a) \geq 0$, and

$$\liminf(\varphi g)(a) \ = \ \varphi(a) \limsup g(a),$$
$$\limsup(\varphi g)(a) \ = \ \varphi(a) \liminf g(a)$$

if $\varphi(a) < 0$.

Proposition A.3 *Let $\mathbf{I}$ and $\mathbf{J}$ be two intervals of $\mathbb{R}$, $\varphi : \mathbf{J} \to \mathbf{I}$ a C^∞-function such that $\varphi'(t) \neq 0$ for every $t \in \mathbf{J}$ and c an interior point of $\mathbf{J}$ such that $a = \varphi(c)$. If $g \in \mathcal{B}_a(\mathbf{I}, \mathbb{R})$ then $g \circ \varphi$ is in $\mathcal{B}_c(\mathbf{J}, \mathbb{R})$ and we have*

$$\liminf g(a) \ = \ \liminf(g \circ \varphi)(c),$$
$$\limsup g(a) \ = \ \limsup(g \circ \varphi)(c).$$

Consider again a real distribution g, defined on the interval $\mathbf{I}$ and bounded near the point a. If we have

$$\liminf g(a) \ = \ \limsup g(a),$$

we will say that g is **continuous** at a and will call the **value** of g at a, denoted by $g(a)$, the common value of its upper and lower limits at that point. The set of all real distributions on $\mathbf{I}$ that are continuous at a will be denoted by $V_a^*(\mathbf{I}, \mathbb{R})$.

---

[4]Remark that a function $G \in \mathcal{C}_a(\mathbf{I}, \mathbb{R})$ has an upper limit at a, denoted by $\overline{G(a)}$, defined in (A.1) and also (when considered as an element of the space $\mathcal{B}_a(\mathbf{I}, \mathbb{R})$) a, possibly different, upper limit at the same point, now denoted by $\limsup G(a)$. The same remark applies to the lower limits, $\underline{G(a)}$ and $\liminf G(a)$. Clearly we shall always have $\underline{G(a)} \leq \liminf G(a) \leq \limsup G(a) \leq \overline{G(a)}$. If, for example, $a = 0$ and $G(x) = \sin\frac{1}{x}$ we have $\underline{G(0)} = -1$, $\overline{G(0)} = 1$ and $\liminf G(0) = \limsup G(0) = 0$.

It is easy to see that every distribution $f \in \mathcal{C}_\infty(\mathbf{I})$ can be represented, in one way only, in the form $f = g + i\,h$, where g and h are real distributions. Thus we say that f is **bounded near the point** a, and we write $f \in \mathcal{B}_a(\mathbf{I})$, if and only if g, h both belong to $\mathcal{B}_a(\mathbf{I}, \mathbb{R})$; f is said to be **continuous** at a if and only if g, h both belong to $V_a^*(\mathbf{I}, \mathbb{R})$. In the last case we write $f \in V_a^*(\mathbf{I})$ and $f(a) = g(a) + i\,h(a)$.

The above definitions of continuity and value of a distribution at a point generalise those introduced in section 2.1. Indeed, a distribution $f = g + i\,h$ (with g, h in $\mathcal{C}_\infty(\mathbf{I}, \mathbb{R})$) is continuous at a (in the sense of this section) if and only if, for sufficiently large values of n, we have

$$g \;=\; \partial_a^n G_n \quad \text{and} \quad h \;=\; \partial_a^n H_n$$

with $G_n, H_n \in \mathcal{C}_a(\mathbf{I}, \mathbb{R})$ such that

$$\lim_{n\to\infty} \overline{G_n(a)} \;=\; \lim_{n\to\infty} \underline{G_n(a)} \quad \text{and} \quad \lim_{n\to\infty} \overline{H_n(a)} \;=\; \lim_{n\to\infty} \underline{H_n(a)}.$$

However, it is easily seen that for the continuity of f in the sense of the section 2.1 we must impose not the last two conditions, but the existence of an integer p such that

$$\overline{G_p(a)} \;=\; \underline{G_p(a)} \quad \text{and} \quad \overline{H_p(a)} \;=\; \underline{H_p(a)} \tag{A.3}$$

(and thus such that $\overline{G_n(a)} = \underline{G_n(a)}$ and $\overline{H_n(a)} = \underline{H_n(a)}$ for all $n \geq p$). It is clear that this last condition is much stronger than the first one.[5]

Consider, as an example, the distribution g defined on $\mathbb{R}$ by

$$g(x) \;=\; \sin \log |x|$$

and the point $a = 0$. Since, for $n \in \mathbb{N}$, we have that $g = \partial_0^{4n} G_{4n}$, where

$$G_{4n}(x) \;=\; \frac{(-1)^n}{4^n}\,\sin \log |x|$$

we obtain immediately,

$$\overline{G_{4n}(0)} \;=\; \frac{1}{4^n} \quad \text{and} \quad \underline{G_{4n}(0)} \;=\; -\,\frac{1}{4^n}$$

and therefore

$$g(0) \;=\; \lim_{n\to\infty} \overline{G_{4n}(0)} \;=\; \lim_{n\to\infty} \underline{G_{4n}(0)} \;=\; 0.$$

However it is easy to see that the distribution g is not continuous in the sense given to this expression in section 2.1.

We may now state some properties of the notions of continuity and value at a point.

---

[5]If we agree to call *"order of the value $g(a)$"* the least integer p for which both equations in (A.3) are satisfied, we could say, in an attempt to make an intuitively acceptable distinction between the two given concepts of value, that one of them is necessarily of finite order while the order of the other may well be infinite.

Proposition A.4 $\mathcal{B}_a(\mathbf{I})$ *is a linear subspace of* $\mathcal{C}_\infty(\mathbf{I})$ *and* $V_a^*(\mathbf{I})$ *is a linear subspace of* $\mathcal{B}_a(\mathbf{I})$. *The map* $f \rightsquigarrow f(a)$, *from* $V_a^*(\mathbf{I})$ *into* $\mathbf{C}$ *is linear.*

Proposition A.5 *If* φ *is a (complex-valued)* C^∞*-function on* $\mathbf{I}$ *and* f *belongs to* $\mathcal{B}_a(\mathbf{I})$, *then* φf *is in* $\mathcal{B}_a(\mathbf{I})$; *moreover, if* $f \in V_a^*(\mathbf{I})$ *then* φf *also belongs to* $V_a^*(\mathbf{I})$ *and* $(\varphi f)(a) = \varphi(a)f(a)$.

Proposition A.6 *Let* $\mathbf{I}$ *and* $\mathbf{J}$ *be two intervals of* $\mathbb{R}$, $\varphi : \mathbf{J} \to \mathbf{I}$ *a* C^∞*-function such that* $\varphi'(t) \neq 0$ *for every* $t \in \mathbf{J}$, c *an interior point of* $\mathbf{J}$ *and* $a = \varphi(c)$. *If* $f \in \mathcal{B}_a(\mathbf{I})$ *then* $f \circ \varphi \in \mathcal{B}_c(\mathbf{J})$; *moreover, if* $f \in V_a^*(\mathbf{I})$ *then* $f \circ \varphi \in V_c^*(\mathbf{J})$ *and* $f(a) = f \circ \varphi(c)$.

We also have a new generalised version of Taylor's formula

Theorem A.7 *Let* a *be an interior point of* $\mathbf{I}$ *and let* f *be a distribution defined on* $\mathbf{I}$ *such that* $\mathbf{D}^n f \in V_a^*(\mathbf{I})$. *Then there exists* $h \in V_a^*(\mathbf{I})$ *such that* $h(a) = \mathbf{D}^n f(a)$ *and*

$$
\begin{aligned}
f(x) \;=\;\; & f(a) + (x - a)\mathbf{D}f(a) + \ldots \\
& + \frac{(x - a)^{n-1}}{(n - 1)!}\,\mathbf{D}^{n-1}f(a) + \frac{(x - a)^n}{n!}\,h(x).
\end{aligned}
$$

As already remarked, every result considered throughout this book, in which the notion of value introduced in section 2.1 has a role, can be generalised to this new setting. The notions of limit, one-sided limit and integral can also be extended in a similar way without any loss of their most relevant properties.

Bibliography

[1] Antosik, P and Mikusinski, J and Sikorski, R (1973), *Theory of distributions, the sequential approach.* Elsevier Publ

[2] Colombeau, J (1984), *New generalized functions and multiplication of distributions.* North-Holland

[3] Ferreira, J C (1965), *Sur la notion de limite d'une distribution à l'infini.* Rend Acc Naz Lincei, 38, 6

[4] Ferreira, J C (1967), *Sur l'intégration de la limite d'une suite de distributions intégrables.* Rend Acc Naz Lincei, 42, 1

[5] Ferreira, J C (1967), *Alguns problemas de prolongamento de distribuições, com aplicações à integração.* Diss Conc Prof Cat, IST - Lisboa

[6] Ferreira, J C (1973), *Sur une notion de limite d'une distribution, plus générale que celle de M S Lojasiewicz.* Rev Fac Ciências de Lisboa

[7] Ferreira, J C (1977), *Sur l'oscillation et la valeur d'une distribution vectorielle en un point.* CMAF - Lisboa, Textos e Notas, 6

[8] Ferreira, J C (1986), *A diferenciabilidade no sentido de Denjoy e a teoria das distribuições.* Mem Ac Ciênc de Lisboa

[9] Ferreira, J C (1987), *Sur une définition de la valeur d'une distribution en un point et son application au problème de Cauchy.* Port Math 44, nº 3

[10] Ferreira, J C (1988), *Uma introdução à axiomática da teoria das distribuições, de J. Sebastião e Silva.* Rev Técnica

[11] Ferreira, J C and Oliveira, J S (1964), *Problèmes au conditions initiales, dans la théorie des distributions.* Rev Fac Ciências de Lisboa

[12] Gel'fand, I and Shilov, G (1964), *Generalized Functions.* Academic Press

[13] Guerreiro, J S (1957), *Les changements de variables en théorie des distributions, I* Port Math, 16

[14] Guerreiro, J S (1959), *La multiplication des distributions comme application linéaire continue.* Port Math, 18

[15] Guerreiro, J S (1961), *Teoria directa das distribuições sobre uma variedade.* Diss Dout Fac Ciências de Lisboa

[16] Guimarães, A A (1959), *Sur une façon de définir, sans dualité, l'espace des distributions tempérés sur la droite et la transformation de Fourier.* Port Math, 18

[17] Guimarães, A A (1968), *Sur une façon de définir, sans dualité, l'espace des distributions vectorielles sur la droite (à valeurs dans un espace localement convexe) et la transformation de Fourier*. An Fac Ciências do Porto, 51

[18] Hoskins, R F and Sousa Pinto, J (1994), *Distributions, ultradistributions and other generalised functions*. Ellis Horwood

[19] König, H (1955), *Multiplikation von distributionen, I*. Math Ann 128

[20] Lighthill, M (1958), *An introduction to Fourier analysis and generalised functions*. Camb Univ Press

[21] Liverman, T (1964), *Generalized functions and direct operational methods*. Prentice-Hall

[22] Lojasiewicz, S (1957), *Sur la valeur et la limite d'une distribution en un point*. Stud Math, 16

[23] Lojasewicz, S (1958), *Sur la fixation de variables dans une distribution*. Stud Math, 17

[24] Lojasewicz, S (1959), *Sur le probléme de la division*. Stud Math, 18

[25] Marinescu, I (1963), *Espaces vectorielles pseudo-topologiques et théorie des distributions*. Deuts Verlag des Wissensch, Berlin

[26] Oliveira, J S (1975), *Sur une méthode de résolution de l'équation $u'(t) + b_0 u(t) + b_1 u(t - \omega) = F(t)$*. Port Math, 34 - 1

[27] Oliveira, J S, Sequeira, F and Ferreira, A V (1977), *Étude de l'équation de Boltzmann du transport de neutrons*. Port Math, 36 - 3/4

[28] Sarrico, C R (1987), *Produtos distribucionais multiplicativos*. Diss Dout Fac Ciências de Lisboa

[29] Schwartz, L (1966), *Théorie des distributions*. Hermann, Paris

[30] Schwartz, L (1945), *Généralization de la notion de fonction, de dérivation, de transformation de Fourier et applications mathématiques et physiques*. Ann Un Grenoble, 21

[31] Schwartz, L (1954), *Sur l'impossibilité de la multiplication de distributions*. Comptes Rendues Acc Sc Paris, 239

[32] Sikorski, R (1961), *Integrals of distributions*. Stud Math, 20

[33] Silva, J S (1954/55), *Sur une construction axiomatique de la théorie des distributions*. Rev Fac Ciências de Lisboa, 4

[34] Silva, J S (1956), *Le calcul opérationel au point de vue des distributions*. Port Math, 14

[35] Silva, J S (1960), *Sur la définition et la structure des distributions vectorielles.* Port Math, 19

[36] Silva, J S (1961), *Sur l'axiomatique des distributions et ses possibles modèles.* Centro Mat Intern Estivo, Florença

[37] Silva, J S (1963), *Novos elementos para a teoria do integral no campo das distribuições.* Mem Ac Ciências de Lisboa

[38] Silva, J S (1967), *Orders of growth and integrals of distributions* (Internacional Course on Distribution Theory). Instituto Gulbenkian de Ciência

[39] Silva, J S (1956), *Introdução à teoria das distribuições.* (Course given at the Centro Estudos de Matemática do Porto published with the text writen by A A Guimarães)

[40] Silva, J S, *Theory of distributions.* (Notes of a Course given at Maryland University which is included in "Obras didácticas de Sebastião e Silva", "Manuais Universitários", Fundação Calouste Gulbenkian),

[41] Sobolev, S (1936), *Méthode nouvelle à résoudre le problème de Cauchy pour les équations hyperboliques normales.* Math Sbornik, 1

[42] Stakgold, I (1967/68), *Boundary value problems of mathematical physics.* MacMillan

[43] Viegas, F S (1987), *Sobre as noções de traço na teoria das distribuições.* Diss Dout IST - Lisboa

[44] Vo-Khac-Khoan (1972), *Distributions, I, II.* Vuibert, Paris

[45] Zemanian (1965), *Distribution theory and transform analysis.* McGraw-Hill

DATE DUE